T0280544

PRECIPITATION

Theory, Measurement and Distribution

Precipitation plays a very significant role in the climate system, every bit as important as temperature, yet this is the first book that provides a comprehensive examination of the processes involved in the growth of clouds, rain, snow and hail. The book also shows how precipitation is measured and what these measurements tell us about its distribution around the globe.

The book begins by tracing our attempts to understand what precipitation is, starting with the ideas of the ancient Greeks, followed, after a long period of stagnation through the Middle Ages, by the improved insights of seventeenth-century scientists, leading to today's detailed knowledge. The measurement of precipitation with rain gauges, radar and satellites is explained followed by what these measurements tell researchers about global annual means, trends, variability and extremes of precipitation. There are still very few rainfall measurements from any of the oceans, while coverage on land is extremely patchy. The concluding chapter outlines what needs to be done to correct these deficits, thereby making it possible to observe future climate change with more confidence.

Precipitation will be invaluable to researchers in environmental science and climatology, those concerned with water resources and flood management, and those planning action as the climate changes. It will also have an appeal to scientific historians.

IAN STRANGEWAYS is Director of TerraData, a consultancy in meteorological and hydrological instrumentation and data collection. He has also written *Measuring the Natural Environment*, now in its second edition (Cambridge University Press 2003).

PRECIPITATION

Theory, Measurement and Distribution

IAN STRANGEWAYS

CAMBRIDGE
UNIVERSITY PRESS

CAMBRIDGE UNIVERSITY PRESS
Cambridge, New York, Melbourne, Madrid, Cape Town, Singapore,
São Paulo, Delhi, Dubai, Tokyo, Mexico City

Cambridge University Press
The Edinburgh Building, Cambridge CB2 8RU, UK

Published in the United States of America by Cambridge University Press, New York

www.cambridge.org
Information on this title: www.cambridge.org/9780521172929

First published 2007
First paperback edition 2010

A catalogue record for this publication is available from the British Library

ISBN 978-0-521-85117-6 Hardback
ISBN 978-0-521-17292-9 Paperback

Contents

Preface

I was able to write most of this book directly from personal experience gained at the Institute of Hydrology (now the Centre for Ecology and Hydrology) in Wallingford from 1964 to 1988 and since then as a consultant in the same disciplines. This whole period has also been regularly interspersed with overseas travel, advising on data collection for water resources, flood warning and agricultural projects. These trips have taken me to about as many remote locations around the globe as David Attenborough has visited in his film making, places rarely seen by outsiders. On these many and diverse missions, from the Antarctic to tropical rainforests and deserts, I came face to face with the complex reality of environmental monitoring and the many problems that obtaining good measurements presents.

When it came to writing Part 4, I needed input from those directly involved in collating, homogenising and analysing the data collected by the global network of raingauges, and more recently by satellites. In this I was helped greatly by David Parker, Jen Hardwick and Chris Folland at the Hadley Centre in Exeter, who supplied me with data and graphs showing long-term precipitation trends in England and Wales, and who helped by passing on some of my questions to others with different specialised knowledge. In consequence I also had a useful and interesting exchange of emails with Aiguo Dai, Kevin Trenberth and Ping Ping Xie all at the National Center for Atmospheric Research in Boulder, Colorado. It is always a pleasure to experience the willingness of people working in science to exchange ideas and to do so with enthusiasm, and all of these co-workers lived up to this expectation.

But before getting to the present day I started by looking back to those times when just a handful of Greek philosophers brought a brief glimpse of light into the pervasive darkness of the ancient world to ponder on the nature of things, including rain and snow and clouds, and at their explanation of what they thought they were – their initiatives only to be extinguished very quickly by the dark ages for the next 2000 years. But then came a renaissance in the seventeenth century and the modern world began, with all its new discoveries; this long history makes interesting reading.

I am grateful to the staff of Cambridge University Press for publishing the book, in particular to Matt Lloyd, who has assisted at every step along the way, and to Dawn Preston, who has managed the production of the book. I would like also in particular to thank my copy-editor, Hugh Brazier, for the way he has been through the book in fine detail, routing out very small, small and not so small errors and suggesting alternative ways of expressing some of the ideas. The result is a book that will be all the more accurate and readable for the close scrutiny it has received.

I should also like to thank the reader for reading the book, and hopefully also for buying it. I can assure you, however, that being an author does not make you rich – unless you write about the adventures of Harry Potter or the private life of a footballer. Fantasy and celebrity pay much better than science. But I would not exchange.

Acronyms and abbreviations

AAO	Antarctic oscillation
ADC	analogue-to-digital conversion
AMS	American Meterological Society
AMSU	advanced microwave sounding unit
AO	Arctic oscillation
ATD	arrival-time difference
ATLAS	autonomous temperature line acquisition system
AVHRR	advanced very high resolution radiometer
AWS	automatic weather station
BT	brightness temperature
CCD	charge-coupled device (image sensor)
CCN	cloud condensation nucleus
CEH	Centre for Ecology and Hydrology (formerly IH)
CERES	clouds and the earth's radiant energy system
CMAP	climate prediction center merged analysis of precipitation
CMIS	conical scanning microwave imager/sounder
CRDF	cathode ray direction finder
CRU	Climatic Research Unit (University of East Anglia)
DCP	data collection platform
DF	direction finder (lightning)
DFIR	double-fence intercomparison reference (snow fence)
DSRT	dual-frequency surface reference technique
DWD	Deutscher Wetterdienst (German National Meteorological Service)
ELF	extra low frequency (radio – lightning)
ENSO	El Niño southern oscillation
ESA	European Space Agency

FRONTIERS	forecasting rain optimised using new techniques of interactively enhanced radar and satellite data
GANDOLF	generating advanced nowcasts for deployment in operational land-surface flood forecasting
GARP	global atmospheric research program
GATE	GARP Atlantic tropical experiment
GCOS	global climate observing system
GEOSS	global earth observation system of systems
GEWEX	global energy and water cycle experiment
GHCN	global historical climatology network
GMI	global precipitation measurement microwave imager
GMS	geostationary meteorological satellite
GOES	geostationary operational environmental satellite
GOMS	geostationary operational meteorological satellite
GPCC	Global Precipitation Climatology Centre
GPCP	global precipitation climatology project
GPI	GOES precipitation index
GPM	global precipitation measurement
GPROF	Goddard profiling algorithm
GPS	global positioning system
GTS	global telecommunication system
ICSU	International Council for Science
IH	Institute of Hydrology (now CEH)
IOC	Intergovernmental Oceanographic Commission
IPCC	Intergovernmental Panel on Climate Change
IR	infrared
JAXA	Japan Aerospace and Exploration Agency
LEO	low-earth orbit
LIS	lightning imaging sensor
LPATS	lightning position and tracking system
MSG-1	Meteosat Second Generation 1
MSS	multispectral scanner
MSU	microwave sounding unit
NAO	North Atlantic oscillation
NASA	National Aeronautical and Space Administration
NCDC	National Climatic Data Center
NOAA	National Oceanographic and Atmospheric Administration
NWS	national weather service
OTD	optical transient detector (lightning)
PIP	precipitation intercomparison project

PIRATA	pilot research moored array in the tropical Atlantic
PR	precipitation radar
RAM	random access memory
RH	relative humidity
RMS	Royal Meteorological Society
RS	remote sensing
SAFIR	(a French lightning direction finding system)
SEVIRI	spinning enhanced visible and infrared imager
SOI	southern oscillation index
SSM/I	special sensor microwave imager
SST	sea surface temperature
SVP	saturation vapour pressure
TAO/TRITON	tropical atmosphere ocean/triangle trans-ocean buoy network
TMI	tropical rainfall measuring mission microwave imager
TOA	time of arrival
TRMM	tropical rainfall measuring mission
UHF	ultra-high frequency
UNEP	United Nations Environment Programme
UV	ultraviolet
VHF	very high frequency
VISSR	visible and IR spin-scan radiometer
VLF	very low frequency (lightning detection)
VP	vapour pressure
WCRP	world climate research programme
WMO	World Meteorological Organization
WV	water vapour

For a fuller list of abbreviations and acronyms, see Padgham, R. C., *A Directory of Acronyms, Abbreviations and Initialisms* (Swindon: Natural Environment Research Council, 1992).

Part 1

Past theories of rain and snow

To understand what precipitation is, it helps to understand the complete hydrological cycle – evaporation, water vapour, convection, condensation, clouds, soil moisture, groundwater and the origin of rivers. To understand these it is necessary to know what the building blocks of matter are, but three thousand years ago none of this was known. In consequence, up to the fifteenth century there were few sound ideas about how the natural world operated. Most of the suggestions were guesses or were based on superstition, religious dogma, legends or myths. Very few observations were made to help form hypotheses and no predictions made that would help confirm a theory. But these ancient views need to be included in a book dealing comprehensively with precipitation, and the first chapter covers the period from 2000 BC up to the seventeenth century AD. The second chapter traces developments over the 300-year period from 1600 to 1900, during which progress accelerated rapidly.

This whole period is also of interest in that it shows how science and rational thought backed by observation, experiment and measurement were rare in antiquity and are still very new to humanity, having started in earnest only a few hundred years ago. This book might, therefore, also be seen as a demonstration and celebration of the progress of the scientific method and of secular thought.

1

The ancients

Ancient Egypt

Around 2000 BC there were four major civilisations: the Egyptian in the Nile Valley, the Sumerian in Mesopotamia (now Iraq), the Harappans in the Indus valley in India, while in China a civilisation grew up on the banks of the Huanghe (Yellow River). None of these people, as far as we know, measured rain or thought about where it came from.

The Egyptians and Sumerians did, however, undertake considerable hydrologic engineering works between 3200 and 600 BC for water supplies, through the construction of dams and irrigation channels (Fig. 1.1). They also undertook the construction of elaborate underground tunnels (*qanāt* systems) for the transport of water over great distances (Biswas 1970).

The ancient Hebrews

The Hebrews in Palestine collected together old stories told verbally for generations, and from them the Old Testament of the Bible grew. While there are numerous references to rain, there is little to suggest that there was any understanding of its cause. Middleton (1965) remarks that most of the Old Testament references to rain merely stressed how welcome it was, and adds that the Hebrews were satisfied simply to marvel at Nature.

The Greeks

On the other hand, the Greeks, close neighbours of the Hebrews, began around 600 BC to question the processes of the natural world and to pursue knowledge for its own sake, rather than for purely practical purposes. Some of the questions they asked were about water, air, clouds and rain, and we need to look at these to put them in a general historical context.

Figure 1.1 This 'falage' flows though a remote village in northern Oman, fed from a spring in the mountainside. It fulfils all the needs of the villagers from domestic water supply to irrigation of the palms. Such falages have been in use for millennia in the arid regions of the Middle East.

Greek views of the elements

Thales, chief of the 'Seven Wise Men' of ancient Greece, believed that the Earth floated on water and that everything came from water, including earth. The first idea may have come from Homer two centuries earlier (eighth century BC) who thought the Earth was surrounded by a vast expanse of water beyond the sea (*Oceanus*), which had no source or origin. This idea was also common in Egyptian and Babylonian cosmology, in which it was thought that the Earth was created out of the primordial water of *Nūn* and that water was still everywhere below it.

The idea that earth arises from water may have been confirmed with some slight logic when Thales visited Egypt and saw the Nile delta being formed by the river – because of the large amount of volcanic silt carried by the river in its annual flood.

The Nile puzzled the Greeks because its flow peaked in the summer while most rivers reach their maximum discharge during the winter. This puzzle was not finally solved until the Victorian explorers Burton, Speke, the Bakers, Livingstone and Stanley ventured into this difficult country during the nineteenth century, suffering many hardships in the process. But there is no time to explore this interesting digression here.

Speculation about the basic elements that everything is made of continued with Heraclitus, who believed it was fire, while Euripides had the view that the elements were air and earth. Empedocles, however, considered fire, air, water and earth to be the elements from which everything is formed, through combination in different proportions. This view became generally accepted later by Plato and Aristotle.

A contemporary of Thales, Anaximander, believed that all life originated in water and that, through continuous evaporation, land emerged from the all-engulfing sea. He also believed that rain was due to moisture being drawn up from the Earth by the sun and that hail is frozen rain, whereas snow is produced when air is trapped in the water. But he would not have called the process 'evaporation' since the idea of water vapour was still a long way off; nevertheless, this was not a bad summary.

Pythagoras developed the idea that the universe could be explained mathematically. None of his books survives but he later had a key influence on Plato's views about the importance of mathematics, which were incorporated into Plato's teachings at the Academy (see later). His view of the significance of numbers was echoed two thousand years later when Johannes Kepler said 'To measure is to know'. This is also a key assertion of this book.

Xenophanes believed that 'the sea is the source of all water, and the source of wind. For neither could come into being without the great main [sea], nor the stream of rivers, nor the showery water of the sky; but the mighty main is the begetter of clouds and winds and rivers' (Freeman 1948). Observing shells and fossil marine animals on high mountains, he suggested that the land must once have been under the sea – and thereby offered some observational proof of his theory.

The Ionian philosopher Anaxagoras appeared to appreciate that rivers depend for their existence on rain (way before his time), but he also believed that rivers depended on 'waters within the earth, for the earth is hollow and has water within its cavities'. Perhaps this was a way of expressing the idea that water was contained within the soil and rocks as soil moisture and groundwater, which is of course perfectly correct, although this is probably not what he meant.

Hippocrates, the father of medicine, thought that water was made of two parts, one thin, light and clear, the other thick, turbid and dark coloured. The sun lifts the lightest part but leaves behind the salty part. Rain, he believed, was the lightest and clearest of all waters. He actually did an experiment by weighing a container of water, leaving it outside and reweighing it after a day. From the drop in weight he concluded that the lightest part had been dissipated; the idea of evaporation and of water vapour was implied but not understood in this test. However, the idea of experiments for conducting scientific investigations had been conceived, although unfortunately this was then forgotten for nearly two millennia.

In his play *The Clouds*, Aristophanes mocked the prevailing idea that rain was sent by the almighty god Zeus:

Strepsiades: No Zeus up aloft in the sky!
               Then, you first must explain, who it is sends the rain:
               Or I really must think you are wrong.
Socrates: Well then, be it known, these send it alone:
               I can prove it by arguments strong.
               Was there ever a shower seen to fall in an hour
               when the sky was all cloudless and blue?
               Yet on a fine day, when the clouds are away,
               he might send one according to you.
Strepsiades: Well, it must be confessed, that chimes in with the rest:
               your words I am forced to believe.
               Yet before, I had dreamed that the rain-water streamed
               from Zeus and his chamber-pot sieve.

Plato was a pupil of Socrates and was greatly affected by the execution of Socrates in 399 BC on the grounds of invented charges of irreverence and of corrupting the young through his teaching and introducing the ideas of atheism, among much else. Following Socrates's execution, Plato gave up his idea of pursuing politics and instead travelled for 14 years through Egypt, Italy and Sicily studying philosophy, geometry, geology, astronomy and religious matters. On his return to Athens in 387 BC, Plato founded the Academy, an institution devoted to research and teaching in philosophy and the sciences, which he presided over in a passionate search for truth until his death.

Plato accepted the idea that there were four basic elements constituting the universe – fire, air, water and earth, as suggested by Empedocles, who had taken the idea from the theology of the Assyrians (in the northern part of ancient Mesopotamia). Plato's main 'scientific' work is his dialogue *The Timaeus*, and in it he reasons that as the world is solid, God must have placed water and air between the extremes of fire and earth. (See 'The views of Aristotle' below for a more elaborate analysis of these four elements.)

He attempts to put this on a mathematical footing using the 'theory of polyhedra' proposed by the mathematician Theaetetos, which says that there can only be five regular solid figures whose faces are regular, identical polygons. Theaetetos was the subject of two of Plato's *Dialogues*, and Plato assigned four of his polyhedra to the four elements (octahedrons to air, icosahedrons to water, tetrahedrons to fire and cubes to earth). The fifth, dodecahedrons, were mysterious and Aristotle later said that Plato must have meant heaven!

Plato had two thoughts on the origin of rivers, the first being similar to that of Anaxagoras – the *Homeric Ocean* – in which there are endless large and small passages inside the Earth with a huge subterranean reservoir *Tartarus* filling the Earth, the waters surging to and fro along the passages, either retiring inside or surging back to fill streams and lakes. This view clearly did not involve rainfall in any way and is probably drawn from myths and poets and priests, expanded by Plato's imagination; it was probably not meant to be the literal truth.

But in the second and more realistic proposal, in the dialogue *The Critias*, Plato says that the rainfall (from Zeus, of course) was not lost to the sea but was absorbed by the soil – 'a natural water pot' – acting as a source for springs and rivers. This is one of the first descriptions of the hydrological cycle involving rainfall and appears also to foresee the idea of soil moisture and groundwater and that rivers flow from these.

The views of Aristotle

A student of Plato at the Academy, and subsequently a teacher there, Aristotle left 20 years later when Plato died and took the job of tutor to Prince Alexander (the Great) for seven years until his accession, thereafter returning to Athens, where he set up the Lyceum. His views of 'the universe' are described in the first three books of his *Meteorologica*. Because of the powerful influence of Aristotle at the time and for the next 2000 years, his views are explained here in some detail.

The elements

Like Plato and probably Socrates and many others such as Empedocles, Aristotle believed that there were four 'earthly' elements, *earth*, *water*, *air* and *fire*, and in his *Meteorologica* Aristotle refines these ideas. There seems to have been some mysterious fifth element (perhaps so that he could include all five polyhedra). Each had two qualities from amongst *hot*, *cold*, *dry* or *humid*.

These elements were not seen as separate things but as being convertible one into another; indeed Thales thought they all originally came from water. Conversion was believed to be induced by *rarefaction* and *condensation* and these depend on heat and cold, but they cannot (according to the theory) be changed in any order but only as shown in Fig. 1.2. For example earth cannot change directly into air, but has to change to water first. (Empedocles believed that it was love and hate that brought about the changes – so presumably nothing changed unless there were people about.) In addition each element had a position relative to the others, so that

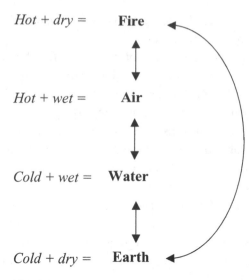

Hot + dry =   **Fire**

Hot + wet =   **Air**

Cold + wet =   **Water**

Cold + dry =   **Earth**

Figure 1.2 Refined by Aristotle, the ancient Greeks believed that everything was composed of four 'elements' – earth, water, air and fire. One could change into another, but only in a certain order, as shown in this diagram. These elements had their own natural positions, fire tending to rise to the top, earth at the bottom, water rising above earth and air above water, but under fire. It was believed that the elements were produced by combinations of *primary qualities* – heat, cold, dryness and moistness, as shown.

fire tended to rise to the top while earth sank to the bottom. Water rose above the earth and air above water, but under fire (Fig. 1.2).

These ideas may sound fanciful now, but one can see how, even through observation if not experiment, this picture of the environment might have been justified. For example clouds can look as if they are appearing 'out of thin air' and a cloud may form over a large fire, the cloud later disappearing, seemingly as if one is changing into the other. They were also aware that the sun evaporated water although they would not have called it that or have understood the process, but it probably made them realise that fire was different to earth, water and air, which are states of matter, whereas fire acts to change the state of the others.

Middleton (1965) explains that the four elements should not be viewed in the same way as our 'elements' of today (hydrogen, helium, lithium . . .), as distinct substances, but rather as 'principles', fire being combustion, not simply a flame, water being fluidity and earth solidity. These ideas were the best that even the cleverest people of the time could come up with, but Aristotle was the first to have made the important step towards something approaching the scientific method, although his failure to make detailed observations or, particularly, to do experiments held back scientific progress until just a few hundred years ago. He believed that all that was necessary was to think about a problem and to discuss ideas, and that the truth would emerge through logic alone. We had to wait 2000 years for that view to be challenged.

Rain, hail, dew, frost and rivers

In the first three books of *Meteorologica* and in several of his other writings, including *Physics* and *On the Heavens*, Aristotle discusses rain, hail, snow, dew, sea, wind, rivers and springs. In all of this he is bound by the generally prevailing view of the four elements, including the idea that air could be converted to water. Despite this he did not agree with the opinion, held by some, that clouds are simply 'thickened' air, and he seems to have had some insight that clouds were composed of water, for he states that the sun's heat evaporates water at the surface and that this rises. When the heat which caused it to rise is lost, the vapour cools and condenses, turning from 'air' into water. Having become water it then falls again onto the earth as rain. From this it is clear that Aristotle did not appreciate the difference between air and water vapour although he knew that 'vapour' came from the action of heat on water; the concepts of evaporation, condensation and water vapour were still 2000 years away.

Aristotle thought that rain was produced in the clouds, but it is not clear how he thought it occurred. He says:

The heat in the clouds rises trying to get free from the clouds, while the cold in the vapour opposes the heat driving the two apart. The cold then presses the particles of the cloud together, combining them into drops, and since water naturally moves downwards the rain falls.

He also believed that dew and frost came from moisture that had evaporated in the daytime but had not risen very far.

Aristotle, along with many others, found it a puzzle why hail, being ice, should occur mostly in the summer, for while he (might have) accepted that small cloud droplets could possibly combine to become bigger rain drops, he could not see how small ice particles could join to form hail, so he concluded that the large drops must have formed before freezing. But he could not then explain how these were suspended high in the air. While Anaxagoras correctly thought that hail was formed when a cloud is forced upwards into a cold region, Aristotle simply contradicted this and said that large hailstones were formed near the ground by 'an intense cause of freezing'. He justified this by saying that the cold is 'concentrated within' by the surrounding heat so that in warmer weather it causes the formation of larger raindrops than in winter and sometimes the concentration of cold by the heat is even greater and freezes the drops, making hail. He introduced the analogy of cool caves on a hot day.

The atomists' view of hail, as stated by Epicurus, observing that hail is always associated with a storm, was that it was formed when a small group of 'water atoms' were surrounded by a large mass of 'wind atoms'. Aristotle and many other later writers denied that wind was air in motion, saying that it was 'a dry exhalation from the earth', although Hippocrates and Anaximander believed that wind was moving air.

Aristotle also used the generally accepted belief that the elements were inter-convertible to explain the origin of rivers, saying that as air can change into water above the ground, so he assumed it did so underground as well. Living in the dry Mediterranean climate, he believed there was not enough rain to produce the flow of rivers and so proclaimed that what happened was that the mountains acted like sponges and soaked up the air, which was converted into water producing the streams and rivers. The origin of rivers had still not been worked out fully even by the eighteenth century.

Theophrastus succeeded Aristotle as head of the Lyceum, but most of his writings are lost, including those on hydrology, although the latter were translated into Syriac and then into Arabic (in 1446). From these it can be seen that his meteorological concepts were the same as those of Aristotle (Stahl 1962). It also seems as if Theophrastus adopted a meteorological explanation of the origin of rivers and was, therefore, probably the first person to have a reasonable understanding of the full hydrological cycle.

The era of Plato, Socrates and Aristotle represented the golden age of ancient philosophy, when considerable intellectual progress was made. It lasted only about 100 years and over the next few hundred years it faded as Rome took the place of Greece as the centre of civilisation.

The Romans

The ancient Greek civilisation started to decline after 300 BC, and by 100 BC the centre of culture had moved to Rome. The Romans were good practical engineers and built impressive aqueducts and good sewage systems but they did this without any obvious design principles or special solutions for construction; it was all a matter of trial-and-error and experience.

Roman writers had very few independent new ideas and simply collected together all the old beliefs from Greece in immense encyclopaedias. A few did try to estab-lish some basic practical principles but generally the Romans were happy with the status quo. In place of the imaginative exploration of ideas in Greece came the standardising of all knowledge in handbooks, in particular those written by Vitru-vius, Seneca, Pliny and Lucretius – enormous compilations of Greek knowledge, mixed in with superstition and travellers' tales. There was nothing new in them, their saving grace being that they transmitted Greek knowledge, albeit imperfectly, to the early Middle Ages. Without them, much of Greek thought would have been lost, and for this reason alone we need to look at the main players in the Roman arena.

One of the great Roman writers, architects and engineers, Marcus Vitruvius, is known mostly through his books *De architectura*, written around 27–23 BC. Book 8 is devoted to water and like his Greek predecessors Vitruvius believed that 'everything proceeds from water'. Vitruvius had read Aristotle and Theophrastus and so had some rough idea of the hydrological cycle. He noted that valleys between mountains had more rainfall than flat areas and that the snow remained on the ground longer because of the dense forests. He appreciated that as the snow melted it percolated down into the soil and eventually produced stream flow. Like Aristotle and Hippocrates he thought that only the lightest and thinnest and more 'wholesome' parts of water evaporated, leaving behind the heaviest, harsh and unpleasant parts. His descriptions are not precise: for example in explaining why there is always more rainfall near mountains than on plains he says that 'Clouds are supported on a *wave of air* and precipitation occurs when they hit mountains because of the shock sustained and because of their fullness and weight.'

His 'bath analogy', however, shows a much better appreciation of the hydrological cycle. In this he says that:

Hot bath water, being heated, vaporises and rising vapour forms droplets on the ceiling. When the droplets become large enough they fall on the head of the bathers. It is reasonable to assume that since there is no source of water on the ceiling, the water must have come from the bath.

Lucius Annaeous Seneca was a Stoic, best known for his moralistic writing and tragedies. His interest to us lies in his *Natural Questions*, written around AD 63, consisting of seven books dealing with astronomy, physics, physical geography and meteorology. But there is nothing new in them, except that they do differ in saying that the wind is moving air – although he then goes on to say that great amounts of air are emitted from the interior of the earth.

Seneca thought that underground water came from:

(1) The earth contains moisture that is forced to the surface.
(2) Air in the earth is converted to water by 'underground forces of perpetual darkness, everlasting cold and inert density'.
(3) Earth within its interior turns itself into water.

He believed that the earth contained subterranean rivers and massive lakes and a hidden sea which produces all rivers on the surface. In this, he says, 'everyone knows that there are some standing waters which have no bottoms and this is the perpetual source of large rivers'.

Gaius Plinius Secundus, known as Pliny the Elder, assembled a massive collection of facts under the title *Natural History*, a huge encyclopaedic collection of past Greek thinking. It was not original and contained nothing new. He upheld Aristotle's view that wind was not air in motion. Despite this he had an untiring curiosity about the natural environment and actually died in the process of observing an eruption of Vesuvius at too close quarters.

A poet and philosopher, Lucretius Carus Titus, whose only known work still to exist is *On the Nature of Things*, seems to have had some realistic concept of the hydrological cycle. Biswas (1970) paraphrases Lucretius's views by saying that 'he believed that moisture rises from everything, especially the sea. When headlong winds drive the clouds across a deep sea, the clouds pick up an abundant supply of moisture just like a woollen fleece soaks up dew. Then the upraised vapours assemble in a thousand ways only to lose their water content for either of two reasons', as expressed by his poem:

> The power of wind drives it along.
> The very multitude of clouds collected in a great array
> And pushing from above,
> Makes rains stream out in copious shower.
> Then too,
> When clouds are scattered by the winds or broken up,
> Smitten above by rays of sun,
> They send their moisture out and drip
> As lighted tapers held above a scorching fire
> Drip fast.

So what came out of the Roman era were many handbooks reiterating Greek thought, but few new ideas.

The Middle Ages

Historians date the end of ancient history and the start of the Middle Ages to the fall of the Roman Empire in the fifth century AD. The next thousand years cover the period known as the Middle Ages or medieval period. The early Middle Ages, from AD 600 to 700, were also known as the Dark Ages. Conditions were indeed bleak and most people were illiterate, clinging to the Christian belief of a better life in heaven. One quotation I think particularly appropriate for this period, although I do not remember the source, is that people were 'saturated in superstition and deluded by hope'.

The power of the church

In the second century AD the schools of Plato and Aristotle were declining and there were now many different religions, sects and philosophies alongside Christianity in Alexandria, which had become the new centre of learning. They did not co-exist very well but one thing united them all: a profound dislike of science.

The Christian Church gradually claimed increasing authority over the human spirit and man was portrayed as helpless and entirely dependent on God for all knowledge. All investigations into natural phenomena were seen as useless, for the truth about everything had been revealed. 'Holy Fathers' such as Tertullian and Augustine opposed any secular enquiry, claiming that scripture was the supreme authority. Knowledge was only allowed as a way of explaining the principles of the scriptures as interpreted by the Holy Fathers. As a result of this suppression, investigations into the natural world faded in Europe and a stagnation spread across all branches of knowledge except for theology. In other parts of the world, notably India, China and the Arab countries there was some progress. During the next thousand years in the West, those who did write on matters that we might consider scientific used them only to interpret the Holy Writ.

Up until the twelfth century, the ideas of the Greek philosophers were known to the Western world entirely through the Roman encyclopaedic handbooks of Vitruvius, Pliny, Seneca and Lucretius. What remained of the original Greek writings were conserved in the East by the Arabs who, from the seventh to twelfth century, made good use of them, especially in mathematics, optics and astronomy, but not particularly in meteorology. The reappearance of these writings in the West in the twelfth century had a strong influence (see 'Aristotle's return', below).

The account flows best through the next thousand years if we look at individual people and what they wrote (Table 1.1). They were relatively few in number.

Saint Isidore

Bishop of Seville, Isidore was an encyclopaedist rather than an original thinker, comparable to Pliny. In his articles *On Clouds* and *On Rain*, in the 20-volume work *Etymologies* or *Origins*, Isidore restates the earlier Greek ideas on meteorology (which he probably got via Pliny and Seneca) but re-interprets them in religious terms. For example he accepts the idea that clouds are condensed or 'contracted' air while rain is described as thickened clouds. When clouds freeze, Isidore says, snow results, and hail he describes as clouds freezing in a more disordered way. But having given a mechanical explanation, Isidore then puts a religious spin on it by saying that clouds are to be understood as holy evangelists who pour the rain of the divine word on those who believe, adding that the air itself is empty and thin,

Table 1.1 *Names of the principal people mentioned in this chapter (in order of date of birth).*

Name	Date of birth and death
Thales	625–548 BC
Anaximander	610–545
Xenophanes	570–470?
Pythagoras	569–475
Heraclitus	535–475
Anaxagoras	500–428
Empedocles	490–430
Euripides	480–406
Socrates	469–399
Hippocrates	460–400?
Aristophanes	448–385?
Plato	427–328
Aristotle	384–322
Theophrastus	371–288
Epicurus	341–270
Lucretius	99–55
Vitruvius	90–20
Seneca	4 BC–AD 64
Pliny the elder	AD 23–79
Isidore (Saint)	570–636
Bede (the Venerable)	673–735
Job (of Edessa)	c.800?
Guillaume de Conches	1080–1154
Michael Scot	1175–1232
Leonardo da Vinci	1452–1519
Antoine Mizauld	1510–1578
Bernard Palissy	1510–?

signifying the empty, wandering minds of men, while the thickened clouds typify the confirmation in the Faith of minds chosen from among the empty vanity of the unfaithful. Hail demonstrates the hardness of treachery and deceit. Snow signifies unbelievers, cold, lazy and dejected.

On water generally, Isidore wrote in Book XIII of *Etymologies* that water was the most powerful element because 'the waters temper the heavens, fertilise the earth, incorporate air in their exhalations, climb aloft and claim the heavens, for what is more marvellous than the waters keeping their place in the heavens' (Brehaut 1912). Although Isidore believed that there was a huge bottomless abyss under the earth from which all springs and rivers flowed and to which they returned through secret channels, like the Platonic concept of Tartarus, he probably got the idea from Ecclesiastes (chapter 1, verse 7), which says:

> All rivers run into the sea;
> Yet the sea is not full;
> Unto the place from whence the rivers come,
> Thither they return.

Isidore gave six reasons for the sea not getting deeper even though rivers flowed into it:

> Its very greatness does not feel the water flowing in.
> The bitter water consumes the fresh that is added.
> The clouds draw up much water to themselves.
> The winds carry it off.
> The sun partly dries it up.
> Because the water leaks through certain secret holes in the earth and turns
>   and runs back to the source of rivers and springs.

This was at least a try. It didn't involve any religious explanation. It even had some small pointers to the truth here and there. Isidore was the greatest medieval encyclopaedist in the West and *Etymologies* was one of the most widely read books for the next thousand years: quite an achievement.

The Venerable Bede

The father of English history, Bede was also the first Englishman to write about the weather and so might be called the founder of English meteorology. A devout churchman living in a monastery in Northumbria, his translation of the Bible was favoured by the Vatican for over a century. With this background it is not surprising that he was only interested in knowledge for propagating church views. But Bede was nevertheless superior to Isidore and along with Boethius can be ranked as one of the most able intellectuals of the Middle Ages in the West. He was not an observer of nature or an original theoriser, but he compiled excellent summaries of existing knowledge from classical sources. His book *De rerum natura* became used as a primer for monks on astronomy and cosmology in which he freely took parts directly from Pliny's *Natural History*, although he always acknowledged this. His views on meteorology were the customary ones from the past and he took the Seneca view that wind was moving air. He pointed out that this could be proved with a fan – a spot of secular thought here.

Iyyūbh or Job of Edessa

Iyyūbh is interesting because of his *Book of Treasures*, a brief encyclopaedia of science written in AD 817. He was clearly aware of Aristotelian views but was sure that wind was moving air and that it was responsible for moving the clouds. But he held the view that rain was due to the inner parts of the clouds turning into continuous liquid water which broke up into drops as it fell. As with all writers of this time he had to give it a religious interpretation by saying that the raindrops were made by a wise God for use by mankind – except for hail, which was to rebuke us for our evil behaviour.

The 'Brethren of Purity' (tenth century AD)

This secret society in Basra wrote an encyclopaedia which shows a considerable advance on Greek thinking about meteorological processes:

If the air is warm, vapours from the sea and from the warmed land rise to a great height, the clouds collecting one above the other, stepwise in spring and autumn. It is as if they were mountains of combed cotton, one over another. But if cold from the zone of ice comes in from above, the vapours collect and become water; then their parts are pressed together and they become drops, increase in weight and fall from the upper region of the cloud down through its mass. These little drops unite with one another until, if they come out of the lower boundary of the cloud, they are large drops of rain. If they meet great cold on the way, they freeze together and become hail before they reach the ground. In consequence, those that come from the upper part of the cloud will be hail, but from the lower boundary of the cloud will be rain mixed with hail . . . So the lower boundary of the region of icy cold, and the high mountains round about the sea confine the two rising streams of vapour from which clouds and rain come; they scatter them and take them away, just like the walls and roofs of bath-houses.

There is no better recorded interpretation of the processes leading to cloud formation and rain production than this until the seventeenth century, and it is also free from theology. The Arabs also made good progress in astronomy and mathematics while the West stood still.

Guillaume de Conches

But in the West, a break was starting to occur between religious and secular views, an example being Guillaume de Conches's down-to-earth mechanical explanation of rainfall:

Rain has various causes. For sometimes thick and wet steam evaporates, and as it ascends, minute droplets become entangled in it. When these have thus become larger and heavier, they fall, and rain occurs. But sometimes air is thickened on account of the coldness of the land and water, and turns into a watery substance which, dried because of the heat of the sun, like ice by fire, falls down in very small particles. Sometimes it happens that the sun, in order to feed its heat, attracts moisture, and the more liquid part of this turns into a fiery substance. This falls down more heavily, when after a very great heat we see a flood of rain. But there is some question about this.

Aristotle's return

For the previous thousand years, the West had known of the work of Aristotle only indirectly through the writings of the Roman encyclopaedists. But the writings of the classical philosophers of Greece had been conserved in the East and now (around AD 1180) re-emerged in the Western world in the form of Latin translations made in Spain from the Arabic versions. The opinions of Aristotle on meteorology, under the title *Meteora,* were again available for all to see and the spirit of enquiry was restarted.

Michael Scot

This Scotsman, a scholar and translator of Aristotle who was also well known as a magician (he even makes an appearance in the eighth circle of Hell in Dante's *Divine Comedy*), wrote asking:

Likewise tell us how it happens that the waters of the sea are sweet although they all come from the living sea. Tell us too, concerning the sweet waters, how they continually gush forth from the earth . . . where they have their source and how it is that certain waters come forth sweet and fresh, some clear, others turbid, others thick and gummy; for we greatly wonder at these things, knowing already that all waters come from the sea, which is the bed and receptacle of all running waters. Hence we would like to know whether there is one place by itself which has sweet water only and one with salt water only, or whether there is one place for both kinds . . . and how the running waters in all parts of the world seem to pour forth of their superabundance continually from their source, although their flow is copious yet they do not increase if more were added beyond the common measure but remain constant at a flow which is uniform or nearly so.

These questions were not answered for about another 500 years, but the break with religion in the search for a rational answer was a good starting point that led eventually to success.

The thirteenth to sixteenth centuries

There were not many significant developments during the thirteenth, fourteenth and fifteenth centuries, all who wrote on the subject of meteorology drawing heavily on Aristotle, and there were no new ideas. One difficulty with the revival of Aristotle's writings was that his idea that wind is not moving air was reintroduced, hampering progress. Also his general philosophy was not helpful to science since he had not advocated experiment or observation. To compound the difficulty further, everyone was afraid to challenge his authority and think for themselves.

Occasionally a view expressed during this time would contain something different, such as in Antoine Mizauld's *Mirror of the Air*, in which he asks why raindrops are round, to which he answers that the shape is good for overcoming the resistance of the air, but then goes on to say that another reason is that all parts of the world, however small, must reflect the round image of the entire universe and so give some sort of example in the form of a pattern or design as far as possible proportional and similar to it. With such ways of thinking, there could be no progress.

The dominance of the church continued during these long concluding centuries of the Middle Ages, only those in the church, law and medicine having a university education. To answer questions about the natural environment the scriptures were consulted, and if they failed to provide an answer the Greek, Roman or Islamic philosophical writings were consulted and dissected and repeated monotonously in a hair-splitting, pedantic manner. New ideas were not proposed and book-learning was preferred to observation, classical literature being the favoured. The Catholic Church and the universities were still hostile to experimental science, but in the fifteenth century came an original new mind.

Leonardo da Vinci

The illegitimate son of a lawyer from Florence, Leonardo was born in the Tuscan village of Vinci in the middle of the fifteenth century. When he was 30 he wrote to the Duke of Milan offering his services as a military engineer and architect and added that he could also do as well as anyone else as a sculptor and painter. He got the job.

He started to write notes from the age of about 37 (in 1489 – three years before Columbus first sailed to the Americas) on scientific subjects, with sketches, but he published none of them. He also wrote in mirror image, being left-handed, and used many obscure abbreviations. When he died, all the notes were given to his friend Francesco Melzi, but he was only interested in the notes about art and tied the rest into bundles which he gave to around 25 libraries and individual people. It was 200 years before anyone bothered to look at them closely but when Giovanni Venturi (1747–1822), the hydraulic engineer, wrote an article in 1797 on them, Leonardo's very modern approach to science was revealed. As a result, his papers were examined in detail and a flood of books and articles followed. Conveniently for us, the greatest number of his notes were on hydrology and hydraulics, providing the first-ever realistic approach to water, rain and rivers; let us see what he said.

In his (never completed) *Treatise on Water*, Leonardo observes that man is made of the same materials as the Earth, comparing bones with rocks as a framework of support, blood and lungs with the oceans and tides. He then speculates that if this was not so it would be impossible for water from the sea to rise up to mountain tops, suggesting that the same thing that keeps blood in our heads keeps water at the summits of mountains through veins in the Earth, and as blood comes out of cuts in our heads, so water issues from cuts in the veins of the Earth.

All of this suggests that Leonardo did not understand the hydrological cycle at all, but he also goes on to make statements that heat (fire) causes a movement of vapour which it raises as mists from the sea up to the cold regions where clouds form, where they become so heavy that they fall as rain. He goes on further to say that if the heat of the sun is added to the power of the element fire, the clouds are driven up higher, to colder regions, where they freeze and then fall as hail.

Now the same heat which holds up so great a weight of water as is seen to rain from the clouds, draws them from below upwards, from the foot of the mountains, and leads and holds them within the summits of the mountains, and these, finding some fissure, issue continuously and cause rivers.

This suggests that Leonardo believed that both accounts were true, the different mechanisms somehow operating together. So while not all of his views about the hydrological cycle were correct, he should not be denied credit for correctly visualising the basic principles of our current view.

Bernard Palissy

Also known as 'the Potter', Palissy, a Frenchman, started out life by making stained-glass windows. He also experimented for years to produce enamelled pottery, which eventually brought him fame and fortune. But in addition he was a keen observer of nature and formed theories based on his observations which put him at variance with authority. To be able to express his ideas in his book *Discours admirables*, published in 1590, he used a similar technique to Plato by presenting them as a dialogue between two people, 'Theory' (the questioner) and 'Practice' (the one who answers).

Using the disguise of 'Practice', Palissy expressed the view that rivers and springs cannot have any other source than rainfall, thereby challenging all philosophers of all times before. He refuted that streams originated from sea water or from air that had turned into water in the ground, saying that the sea would have to be higher than the mountains and rivers would be salty. He did not believe that air could turn into water in the ground and so did not cause rivers, and accepted that water could form by the condensation of vapour, although he said there would not be enough to cause rivers directly. Palissy went on to state with confidence that (slightly edited):

The action of the sun and dry winds, striking the land, causes great quantities of water to rise, which being gathered in the air and formed into clouds, have gone in all directions like heralds sent by God. And when the winds push these vapours the waters fall on all parts of the land, and when it pleases God that these clouds (which are nothing more than a mass of water) should dissolve, these vapours are turned into rain that falls on the ground.

He then goes on to describe his view of how rivers form:

And these (rain) waters, falling on these mountains through the ground and cracks, always descend and do not stop until they find some region blocked by stones or rock very close-set and condensed. And then they rest on such a bottom and having found some channel or other opening, they flow out as fountains or brooks or rivers according to the size of the opening and receptacles; and since such a spring cannot throw itself (against nature) on the mountains, it descends into valleys. And even though the beginnings of such springs coming from the mountains are not very large, they receive aid from all sides, to enlarge and augment them; and particularly from the lands and mountains to the right and left of these springs.

This, said Palissy, is the correct explanation and no one need look further. Unfortunately, because he wrote in French when most scholars at that time wrote (and read) in Latin or Greek, his views went largely unknown, as were Leonardo's because he never published any of them. So despite these two men, little changed in our general understanding for quite a while yet. But better was to come – eventually – after a hard struggle – as the next chapter reveals.

References

Biswas, A. K. (1970). *History of Hydrology*. London: North-Holland.

Brehaut, E. (1912). *An Encyclopedist of the Dark Ages: Isidore of Seville*. Studies in History, Economics and Public Law. New York, NY: Columbia University.

Freeman, K. (1948). *Ancilla to the Pre-Socratic Philosophers*. (Translation of H. Diels, *Die Fragmente der Vorsokratiker*, 5th edn.) Oxford: Blackwell, p. 23.

Middleton, W. E. K. (1965). *A History of the Theories of Rain and Other Forms of Precipitation*. London: Oldbourne.

Stahl, W. H. (1962). *Roman Science*. Madison, WI: University of Wisconsin Press.

2

A renaissance

In many ways the path to understanding something like rain or evaporation is more interesting than the final discovery. The struggle of thinkers to comprehend them is as intriguing as the final facts that eventually emerged. This chapter well illustrates the difficulties and the slow dawning. The same is certainly going on today in cosmology and quantum physics, and probably in climatology too. It is difficult to say exactly who or what started the revolution of thought that occurred in the seventeenth century, but much suddenly happened all at once.

The 'new philosophy' – Empiricism

England's greatest contribution to philosophy was probably *Empiricism*, a pragmatic philosophy, practical, hard-headed, no-nonsense, Anglo-Saxon, the exact opposite of rhetoric. Empiricism also had a strong association with British Protestantism, a movement that aimed to demystify the church itself and also the monarchy, which was still mystical and closely related to religious dogma. The Protestants said the monarchy had no divine right and was not somehow speaking for God: they were just ordinary people. Francis Bacon was one of the first to promote this line of thinking.

Although primarily a politician (appointed to England's high office of Lord Chancellor in 1621, five years after the death of Shakespeare) Francis Bacon's greatest interest lay in the search for scientific truth. Up until the seventeenth century, as shown in the previous chapter, science (if it could yet be called that) was (in the West) based on Aristotle's view that any truth could be reached simply by thinking, by argument and by discussion. ('Thought experiments' may be an exception.) Bacon rejected this, saying that all scholarly activity in the universities had become nothing more than learning what the Greeks had said by rote, with hair-splitting rhetorical discussion. He said that progress in the new sciences was being stifled. While Bacon did not contribute specifically to any increased understanding of

21

matters such as precipitation, his scientific method laid the foundation for a new way to investigate it, and was indeed the basis for all scientific enquiries. 'Whether or no anything can be known, can be settled not by arguing but by trying', he said in his book *Novum Organum* (1620).

The Royal Society set itself up in 1662 as a Baconian institution, while in France the Académie Royale des Sciences was formed in 1666. Unfortunately, despite Bacon's forward-looking view, he constantly referred back to Pliny and still believed that wind was produced by 'exhalations of the earth' and by 'vapours raised by the heat of the sun', such vapours turning into wind or into rain. 'Winds do contract themselves into rain', he said, not admitting that he did not know how rain was formed.

John Locke (1690) was Empiricism's defining philosopher. Following the example Bacon had set a few years earlier, Locke argued that scientific truth needs evidence from the real world in the form of controlled experiments and observations; at this time 'experiment' and 'experience' meant more or less the same thing. By a process of *induction*, general truths about the world can then be arrived at – the *scientific method*.

Although Locke respected Aristotle's enormous philosophical contributions in all other fields, at the heart of Locke's approach was an extreme dislike of Aristotle's un-scientific methods. Apart from the reasons already given for this, Locke also objected to the way in which Aristotle divided studies into isolated compartments – animals, machines, heavenly matters – considering them separately and differently. The new philosophy aimed for a unified view of them all, in one continuum, with common explanations. But we cannot be too critical. Aristotle was amongst the first people ever to think objectively on a large scale. The problem was that people lived in such awe of him for millennia that progress was difficult. No one dared to question what he had said. So it was the timidity of others, rather than Aristotle's authority, which was the problem. In the seventeenth century there were many minds at work in Europe, buoyed by the new-found freedom of thought engendered by Empiricism, each stimulating the other. Locke was particularly influenced, for example, by René Descartes.

This chapter is divided into three parts, dealing with early ideas about water vapour, about clouds and about precipitation (rain, snow and hail). Most of the people mentioned in the chapter are listed in Table 2.1.

Emerging ideas about water vapour

The process of evaporation and the nature of the resulting water vapour, how it integrates into the atmosphere and how it can then reappear as liquid water was a puzzle to thinkers at the start of the seventeenth century. The twists and turns are

Table 2.1 *Names of the principal people mentioned in this chapter (in order of date of birth).*

Name	Dates of birth and death
Francis Bacon	1561–1626
René Descartes	1596–1650
Otto von Guericke	1602–1686
Robert Boyle	1627–1691
John Locke	1632–1704
Edward Barlow	1639–1719
Philippe De la Hire	1640–1718
Isaac Newton	1642–1727
Gottfried Wilhelm Leibniz	1646–1716
Edmond Halley	1656–1742
William Derham	1657–1725
Thomas Molyneux	1661–1733
Guillaume Amontons	1663–1705
Thomas Newcomen	1663–1729
John Theophilus Desaguliers	1683–1744
Henry Beighton	1687–1743
Jean Bouillet	1690–1777
Pieter van Musschenbroek	1692–1761
Charles François de Cisternay Du Fay	1698–1739
Daniel Bernoulli	1700–1782
Georg Wolfgang Krafft	1701–1754
John Rowning	1701–1771
Nils Wallerius	1706–1764
Benjamin Franklin	1706–1790
William Cullen	1710–1790
William Heberden	1710–1801
Christian Gottlieb Kratzenstein	1723–1795
Charles Le Roy	1726–1779
Daines Barrington	1727–1800
Jean André Deluc	1727–1817
Johann Heinrich Lambert	1728–1777
Hugh Hamilton	1729–1805
James Six	1731–1793
Erasmus Darwin	1731–1802
Joseph Priestly	1733–1804
James Watt	1736–1819
Marcellin Ducarla-Bonifas (Du Carla)	1738–1816
Horace Bénédict de Saussure	1740–1799
Antoine Laurent de Lavoisier	1743–1794
Gaspard Monge	1746–1818
Jacques Charles	1746–1823
Marc-Auguste Pictet	1752–1825
Benjamin Thompson (Count Rumford)	1753–1814
William Charles Wells	1757–1817
John Gough	1757–1825

(cont.)

Table 2.1 (*cont.*)

Name	Dates of birth and death
Patrick Wilson	1758–1788
John Leslie	1766–1832
John Dalton	1766–1844
Luke Howard	1772–1864
Robert Brown	1773–1858
Leopold von Buch	1774–1853
Angelo Bellani	1776–1852
Humphry Davy	1778–1829
Joseph Louis Gay-Lussac	1778–1850
Siméon Denis Poisson	1781–1840
James Pollard Espy	1785–1860
John Herepath	1790–1868
Macedonio Melloni	1798–1854
George Biddell Airy	1801–1892
Heinrich Gustav Magnus	1802–1870
Pierre Hermand Maille	1802–1882
James Glaisher	1809–1903
Henri Victor Regnault	1810–1878
Auguste Bravais	1811–1863
Elias Loomis	1811–1889
Augustus Waller	1816–1870
Joseph Dalton Hooker	1817–1911
Paul-Jean Coulier	1824–1890
William Thomson (Baron Kelvin)	1824–1907
William Stanley Jevons	1835–1882
Julius von Wiesner	1838–1916
John Aitken	1839–1919
Osborne Reynolds	1842–1912
Phillip Lenard	1862–1947
Wilhelm Matthaus Schmidt	1883–1936
Albert Defant	1884–1974

rather confusing but that makes it all the more interesting. What follows is not in strict chronological order but is presented in a way that flows most easily.

Water does not turn into air

Descartes (1637) believed that all materials were made of particles of the same substance but of different sizes, all in motion and separated by a different material that he called 'subtle matter'. Water, by his theory, was made of long, smooth, eel-shaped particles that could be separated without difficulty, while air and liquids differed from solids only in as much as the particles of solids were closer together and interwoven. Heat, he believed, was an expression of the subtle matter in motion, the higher the temperature the faster the movement. The freezing of water was due to a lesser movement of the subtle matter, which could no longer keep the eel-like particles apart.

Using this view of matter, Descartes explained the evaporation of water by saying that as the subtle matter becomes more agitated by the heat of the sun (or some other heat source) it makes the water particles move more strongly and as a result some break away from the water and fly off into the air – as dust rises when disturbed. Descartes emphasises that this is not because they have a predisposition to go upwards but because there is less resistance in the air to the motion.

At this time everything that we now call 'gas' was called 'air'. For this reason it was assumed that water *became* air when it was heated and evaporated. Descartes's idea that water *stayed as water* but became a vapour when it evaporated was, therefore, an important intellectual insight. The idea of water being particles in motion was also ahead of its time, foreseeing the kinetic theory of gases, which came much later.

Along similar lines of thought, Otto von Guericke observed that on a summer's day, wine containers, brought from the cold of a cellar, sweat, the reason being, he believed, that the air in contact with the vessel is cooled and is thereby contracted, squeezing the water out. Although the actual mechanism postulated was not correct, the idea that water was converted to vapour, not into air, when heated, and that the process could be reversed by cooling, was gaining credibility.

Vapour visualised as minute bubbles

Descartes's analogy with rising dust was not helpful since eventually dust settles back down to the ground again, but in 1666 Urbano d'Aviso, an obscure cleric, suggested that water vapour was composed of minute bubbles of water filled with fire, which rise through the air so long as the air is of greater specific gravity than the bubbles. When they arrive at a height where they are the same density as the air, they stop.

In 1688 Edmond Halley gave an explanation to the Royal Society, seemingly similar to that of Urbano d'Aviso but apparently arrived at independently, although he was not sure what was in the bubbles, thinking it might be some sort of matter that countered gravity; he cited the fact that plant shoots grow upward for no obvious reason, and likened the two.

Water vapour is lighter than air

In his *Opticks*, Newton (1704) said emphatically that water vapour might rise into the air if it was lighter than air, adding that a moist atmosphere would then be lighter than a dry one. This was not well received because the view held at the time was that if vapour (indeed if anything) was added to the air already there, the mixture would *inevitably* become heavier. Newton had no experimental proof of his claim but seemed to see the answer intuitively. (Intuition is not of course always a good tool in science, which often turns out to be counter-intuitive.) Others were also having their say on water vapour from quite different angles.

The properties of steam

The Industrial Revolution was under way in England and the properties of steam, important for industrial power, began to influence the evolution of thought about water vapour. Here the work of Thomas Newcomen, who had made the first practical steam engine using a piston in a cylinder, was helpful. His 'atmospheric engine' was designed to operate a water pump in Cornish tin mines to stop flooding, a reduced pressure being formed in the cylinder by condensing steam, atmospheric pressure then forcing the piston up the cylinder.

Following up on Newcomen's developments, the physicist John Theophilus Desaguliers and a surveyor, Henry Beighton (1729), investigated the processes involved in the generation of steam and from their measurements concluded, incorrectly, that when water is boiled it expands to 14 000 times its liquid volume when it is converted into steam (the correct value is around 1670), while water heated from 0 °C to 100 °C (but not converted to steam) expands by 1/26 of its volume (3.85%). Yet again, air heated from 0 °C to 100 °C expands by 5/3 of its original volume (166.7%).

The concept of *latent heat of vaporisation* was not known at the time and consequently it was not understood that much more energy is required to convert water at boiling point into steam at boiling point than is required to raise its temperature from freezing to boiling point. But the real importance of their work was that they seemed to grasp that the invisible water vapour just above the surface of the boiling water, before it turns rapidly into visible 'steam' and then rapidly disappears again, is the very same invisible vapour that arises (evaporates) from water at a lower temperature. (*Steam* is invisible water vapour; the visible cloud of minute water droplets is not steam but 'cloud', even though we often, colloquially, still call it steam.) Another important statement in the paper was that increasing the repellent force of the particles of an inelastic (incompressible) fluid makes it elastic and vice versa and that heat increases the repellent force, again suggesting some increasing appreciation of the evaporation process and of the kinetic theory of gases.

Water particles expand, balloon-like

But despite these dawning new insights, confusion continued elsewhere. In a much-read physics textbook of the time, Pieter (Petrus) van Musschenbroek (1736) considered the case of boiling water and hypothesised that the heat 'insinuates' itself into the 'little pores' of each water 'particle', causing them to expand greatly (balloon-like) by (the above quoted) 14 000 times, whereupon they rise until reaching equilibrium with the air. He wondered how something could expand so much without 'bursting', but had no answer. To add further to the uncertainty, Musschenbroek lumped into his considerations such things as the spray at the base of high waterfalls, adding to his vagueness about the difference between invisible water vapour and visible 'steam' or the visible spray of the waterfall that looked like the 'steam' from a kettle.

The 'solution theory' of water vapour

Earlier, Thomas Molyneux, aided by his brother William, published a paper on why heavy materials, dissolved in liquids lighter than they are, stay dissolved and remain suspended indefinitely (Molyneux 1686). It was suggested that the particles of the dissolved material were so small that they can be moved by the merest of forces, the movement of the solute particles being sufficient to drive them from place to place and so overcome the force of gravity. Although Halley also commented on the similarity between dissolved solids in water and water vapour in air, the link was not developed further in the seventeenth century.

Then Jean Bouillet (1742) proposed that the small particles of water were in rapid motion at the surface of liquid water and would fly off into the air (a preview of Brownian motion). But he was puzzled why they did not just fall back (water being heavier than air). To account for this he proposed that the water 'united' with the air. Although not so called, this is in effect a 'solution' theory.

There then came a very important paper by Charles Le Roy (1751) of Montpellier, in which he suggested that the 'particles of water' were 'supported' in the air by a process similar to the mechanics by which solids dissolve in liquids. To demonstrate the effect of temperature on this process and to show the similarity, Le Roy sealed some damp air in a bottle and cooled the bottle, dew forming inside, which disappeared when the bottle was warmed up again. Dew always appeared at the same temperature – the lowest temperature, Le Roy thought, that the air could hold 'in solution' the water that was sealed in the bottle. He compared this with how a solution, if cooled, could not continue to hold all the salt dissolved in it which then crystallised out. We would now call this the *dew point* but at the time Le Roy called it *le degré de saturation de l'air*.

He also measured the dew point of air in a room and outside by putting ice in a glass of water, pouring the cooled water into a dry 'goblet'. If dew formed he waited until it warmed half a degree and then poured it into another dry glass, and so on until dew just did not form. He made numerous observations using this arrangement, such as observing that the dew point of the atmosphere changed with wind direction. He saw the relevance of this to ideas about the suspension of water in air as well as to the source of precipitation. He even went on to see that existing ideas about 'moist' and 'dry' air do not refer to the absolute water content but to the maximum amount the air can contain at any one temperature – 'dry' air on a warm day can contain more water than 'moist' air on a cold day. The idea of relative humidity had been created. His paper was a very important contribution to meteorological thinking in the eighteenth century.

Others championed the idea of 'solution' to explain how water vapour entered the air, one being the Reverend Hugh Hamilton, Professor of Natural Philosophy at Dublin. He presented a paper pointing out additional similarities between evaporation and the solution of solids in liquids (Hamilton 1765): both are speeded up by stirring/wind; heat encourages, while cold slows down the processes; dew is the equivalent to crystals forming as the air/solute cools; the evaporation of liquids and the solution of salt both result in cooling. But when Hamilton presented his views, the similarity was noted with a similar paper submitted by Benjamin Franklin (1765) nine years before – which was then published with apologies.

The Swiss physicist Horace Bénédict de Saussure, inventor of the hair hygrograph, developed the solution theory further in his *Essais sur l'hygrométrie* (1783) and said that evaporation is produced by an interaction between fire and water which produces an elastic fluid, lighter than air, which he called *vapour*. When the vapour is produced in a vacuum or by boiling water, he referred to it as *pure elastic vapour*, while when it is formed at room temperature in the air, the vapour mixes with it and goes into true solution, to which de Saussure gave the name *dissolved elastic vapour*. This differs from previous speculations in that it requires the combination of water with fire before it 'dissolves'.

But the 'solution' idea had a serious flaw. Why does a liquid evaporate in a vacuum? This had been shown to happen in earlier experiments with air pumps when Nils Wallerius (1738) published a paper in which he reported finding that water did evaporate in a vacuum and so must rise up into it. Although this paper remained unknown, it would have shown, if anyone had noticed it, that the explanation for the ascent of the vapour could not be because it was lighter than air, because it could not be lighter than a vacuum and also nothing could 'dissolve' in empty space. Various people tried to explain this away or even denied it happened. One such was a Dr D. Dobson of Liverpool, who in 1777 did an experiment in which he 'weighed two china saucers, each containing three ounces of water.' He placed one out of doors and one under the receiver of an air pump. He claimed that after four hours the saucer outside had lost weight while the weight of the one in the vacuum 'was not sensibly diminished'. As Middleton (1965) remarks, it is difficult to believe that

Dobson was both honest and competent. We need not waste time on these evasions, but they do show what a struggle it was getting to the truth.

A combination of water and fire

In the eighteenth century, fire was seen as a vague material, called 'caloric', Jean André Deluc being a prominent advocate of this view. By 1760 he was convinced that water vapour is lighter than air because it is composed of a mixture of water and fire. Lavoisier agreed and wrote that fire was a subtle fluid material thing that he called *igneous fluid*, which penetrates all substances but not in equal amounts, and that it could also exists on its own. Temperature is an indicator of how much of this fluid is free. By measuring the change in temperature in a reaction we can tell if *matière du feu* is generated or absorbed. Using this logic, Lavoisier concluded that as evaporation causes cooling, heat must be combining with the fluid to produce the vapour. (Middleton (1965) asks where Lavoisier would have said the heat came from to cause evaporation if the liquid was poured out in the space between the earth and moon.)

Deluc accepted this description and went on in 1786 to form a hypothesis to explain the mechanism of evaporation by suggesting that air or water vapour or any gas (*expansible fluids* as he called them) was made up of discrete moving particles and that the pressure exerted by the water vapour was independent of any air that might be mixed with it. At any one temperature it is the same in a vacuum as when mixed with air. This was yet another glimpse of the idea of the kinetic theory of gases, but it lacked the observational proof to back it up. This followed 30 years later.

The barometric pressure enigma

Another quite different approach to the question of water vapour in the atmosphere centred on the newly invented mercury barometer. Torricelli made the first barometer in 1643/44 and they soon became used in an attempt to understand the behaviour of the atmosphere and weather.

The first problem was that the highest pressures occurred in fine weather, the lowest on rainy days – when the air is 'full of vapours and exhalations' and, thus, 'obviously heavier'. Many unrealistic ideas were proposed to account for this, such as the rain and clouds 'sustaining' part of the weight of the atmosphere above them, although how this might happen was not discussed.

Another problem with understanding changes in barometric pressure hinged on the fact that it was believed that the atmosphere had a definite upper boundary, like the sea, at the same altitude all round the globe. A suggestion that the height of the atmosphere might vary was made in 1705 by Philippe de la Hire at the Paris Observatory when he hypothesised that the variations of the barometer corresponded to changes in the depth of the atmosphere.

Soon afterwards, however, the work of Guillaume Amontons on thermometry showed that temperature had quite a large effect on the density of air, causing Placentini to propose that the main cause of barometric pressure changes might be due to variations of atmospheric temperature. He argued that as the height of a column of atmosphere remains constant, being a fluid, its weight and thus barometric pressure will vary as the density varies (with temperature). This implied that air must move sideways (advect) as it expands or contracts (with changes of temperature) so as to maintain its level.

It was also argued that if water vapour was added to the air, by evaporation, its extra mass would increase the barometric pressure. But Deluc's view was that as the added vapour could not increase the height of the atmosphere, which must remain constant, the extra volume 'spills over' into the adjacent air, and as the vapour is lighter than air, this would cause the pressure to drop. But then de Saussure did an experiment to investigate this using a large glass globe containing saturated air at 16 °C and found that the removal of all the water vapour reduced the pressure by only around 2 per cent. He concluded that even if the atmosphere passed from dry to saturated, the change of barometric pressure would be much smaller than is experienced in practice and so changes in temperature, much more than water vapour, must be the main cause of variation in barometric pressure. John Dalton held the same views. They were of course absolutely correct.

Electricity and chemistry: an unfruitful diversion

Between 1730 and 1760 knowledge of static electricity had grown quickly and like all new ideas it got drawn into meteorological debates, although not profitably, to try to explain the processes of evaporation and the rise of water vapour into the atmosphere.

Likewise, towards the end of the eighteenth century, particularly in the 1780s, huge advances were made in chemistry by Black, Cavendish, Priestly, Scheele, Lavoisier and others. What is relevant to us here was the discovery that air is a mixture (mostly) of two gases – now called oxygen and nitrogen – rather than a simple 'elementary' substance. Also discovered was that water was a compound of oxygen and hydrogen. With air and water being the major constituents of the meteorological environment, these discoveries were invaluable. But chemistry got drawn into the debate, just as static electricity had, to explain evaporation and condensation, again without any benefit – and we will not spend time on this sidetrack.

A static theory of gases

In 1793 John Dalton expressed doubts concerning chemistry's involvement in basic meteorological processes, saying that evaporation and condensation of water vapour do not involve chemical reactions of any sort and that aqueous vapour always exists as a fluid diffused in amongst the rest of the 'aerial fluids'. Dalton was now a significant player, and in 1801 he gave a paper in Manchester in which he stated several laws that he had derived from his experimental work (Dalton 1802).

In the *law of thermal dilation*, supported by Gay-Lussac's experiments in the same year, Dalton stated that all gases have the same expansion if heated by the same amount. But fourteen years earlier in 1787 Jacques Charles had done similar experiments in France (using oxygen, nitrogen, hydrogen, carbon dioxide and air) and for this reason the thermal dilation of gases is usually known as Charles's law. This is examined along with Boyle's law in the next chapter. (The problem was that Charles had not bothered to publish anything on it.)

In the *law of partial pressures*, Dalton stated that when two gases, A and B, are mixed in a container, there is *mutual repulsion* between all the particles of gas A and between all the particles of gas B, but no repulsion between any of the gas A and gas B particles. In effect, gas A and gas B act as if they were alone in the space: 'One gas is as a vacuum to another.' Additionally, Dalton says, the space occupied by a gas is little more than a vacuum because the gases are very tenuous. Consequently, Dalton concluded, the pressure (or total weight) of the combination of gases is the sum of the two acting as if alone. While this law of partial pressures was correct, Dalton was incorrect in assuming that the explanation was that the gas particles repel each other. They do not – all the gas 'particles' (type A and B) attract each other (when very close together – see Chapter 3). Dalton's view can, therefore, be seen as a 'static' theory of gases based on the incorrect assumption that the gas particles fill all the space available because they repel each other.

The kinetic theory of gases

Returning to the mid eighteenth century, Daniel Bernoulli, in his *Hydrodynamica* (1738), proposed that the relationship between the pressure, temperature and volume of a gas, developed by Boyle a century earlier (more details in the next chapter), could be explained by assuming that the gas was composed of numerous tiny particles *in rapid motion*. This, however, was not accepted at the time.

In 1786 Joseph Priestly said that when any two gases ('airs') of different specific gravities, known not to have any cause of attraction to each other, are mixed together in a vessel, they are never separated again by mere rest, but remain diffused equally through each other.

In the same year, Deluc wrote concerning the relationship between the maximum amount of water vapour that air can hold and the temperature of the air (see also Fig. 3.1). He went on to say that this maximum was the same whether or not the space contained air or a vacuum. This he showed by experiment. It thus follows, he said, that the minimum distance of the water vapour particles (which determines the maximum density) involves only the vapour itself and is independent of any other gases present.

If he (or anyone else) had paid attention to Bernoulli's *Hydrodynamica*, they would also have seen that the explanation as to how gases fully intermingle, whatever their specific gravity, and fill the full space available to them, lay in the 'kinetic' theory of gases, that is that the particles are all in continuous and rapid motion. Dalton's mistake over this (assuming that the particles repelled each other) led to his not understanding the process of evaporation and water vapour in the atmosphere correctly, and so he went on to argue that if all the other gases (the air) were removed 'little addition would be made to the aqueous atmosphere, because it already exists in every place, almost entirely up to what the temperature will admit'. This implied that the atmosphere is already nearly saturated everywhere. 'What then,' asks Middleton (1994), 'constitutes dry weather or wet weather?'

The proportion of oxygen to nitrogen in the atmosphere

An important question that arose when considering how the different gases that constitute the atmosphere mix is how the proportions of oxygen and nitrogen vary with altitude. There still lingered the thought, before the kinetic theory had been derived, that a heavier gas would 'settle out' beneath the lighter, barring any mixing effects induced by wind. The very risky balloon flight undertaken by Gay-Lussac in 1804 to an altitude of around 20 000 feet (6100 m), taking air samples at various heights, helped to answer this. It was found that oxygen always accounted for 21.49% of the air whatever height the sample was taken at. The atmosphere was shown to be *thoroughly mixed*. Welsh and Glaisher also made balloon ascents and from these derived an empirical relationship between vapour pressure and altitude. The tests showed, incidentally, that 90% of all water vapour in the atmosphere is below 20 000 feet. Another expedition to high altitude was made by Joseph Hooker (of Kew Garden fame) in the Himalayas, where he showed that the vapour pressure at 18 000 feet (5500 m) was only a quarter of that which Dalton had predicted.

In parallel with these great outdoor adventures, physicists and mathematicians, safely at work in labs and offices, determined with increasing precision the maximum vapour pressure at different temperatures. In all of these, the presence of air with the water vapour was supposed to have no effect, but in 1854 Henri Victor

Regnault showed that in all of his tests, amounting to around 90, the saturation vapour pressure was (a *soupçon*) less in air than in a vacuum.

Brownian movement

Between 1816 and 1821 John Herepath published papers that derived the gas laws and gave explanations of diffusion, changes of state and how sound travels, based on a kinetic theory of gases. This was developed in complete independence of a similar idea first proposed by Bernouilli 50 years earlier which had been ignored. Herepath's was also not given a fair airing (through personality clashes).

Then in 1827 Robert Brown, a Scot, while using a microscope to make biological observations concerning the fertilisation of flowering plants, noticed that pollen grains were 'filled with particles' that were 'very evidently in motion'. Note that it was not the pollen grains, as is so often reported, that were vibrating, but much smaller grains (about 2 μm diameter) within the pollen. Others had seen this motion earlier and had concluded that it was some sign of life, but having seen the motion in many living specimens Brown wondered if it persisted in dead plants. Indeed it did, for he saw the same movement in pollen grains preserved in an alcoholic solution for 11 months. He went on to observe the motion in finely powdered rock and concluded that any solid mineral would show the movement if in a fine enough powdery state. He concluded that the jiggling motion was due to the surrounding molecules striking the particles at high speed.

What water vapour is and why it fills the full space available to it was now solved, but another problem that had to be tackled was the composition of clouds and what caused them to remain suspended. To follow developments on this front we must again go back to Descartes.

Emerging ideas about clouds
What keeps clouds aloft?

A concern everyone had in the early seventeenth century was why clouds did not fall, considering that water is heavier than air, and clouds were known to be made of water in some form. A related preoccupation was whether clouds were formed of bubbles or drops.

It is worth going back briefly to note that Aristotle had said that clouds were *drops of water* that ride aloft because their smallness allows them to rest on the air. This was near the truth, but was forgotten or ignored.

Jumping two thousand years, Descartes explained clouds by saying 'As the particles of the cloud have a large surface in relation to the quantity of matter in them, the resistance of the air that they would have to divide if they were to descend can easily have more force to hinder them than their weight has to drive them down.' He also suggested that the wind not only raised the clouds but also sustained them. This is as near to the truth as anyone was to come for the next hundred years, but he got many other things wrong, one being that he saw clouds as self-contained entities, like balloons, that were pushed around as a unit by the wind.

Are cloud particles vesicular?

But these near-correct views did not prevail, and the most common belief held during the early eighteenth century, to account for why clouds floated and did not fall, was that cloud droplets were not drops but vesicular, that is were composed of tiny *bubbles* of water with air, or perhaps something lighter, inside them, like those blown with soapy water.

The earliest suggestion that clouds were formed of tiny bubbles came in 1701 from a Jesuit, Pardies, who supposed that clouds were composed of 'fiery spirits' or 'rarefied air'. He made this statement while trying to contradict a view held at this time that clouds were held together in some way 'as if by glue' to form a single body. He said he doubted that clouds were any more firmly bound as a unit than water, which allows air bubbles to pass easily through.

Otto von Guericke, using his air pump, produced clouds artificially in something akin to a modern cloud chamber, the pressure suddenly being reduced in a container of moist air. On days when there were just a small number of nuclei in the air (anticipating Chapter 5), a few large drops formed and sank quickly to the bottom of the chamber, while on days when there were many nuclei, there were more and smaller drops that stayed suspended longer. Guericke was unaware of the need for nuclei to form cloud droplets and seems to have concluded from his observations that two different processes were involved, the smaller particles being hollow while the larger ones were drops. His reasons are not clear, but Middleton (1965) believes that this was indeed his conclusion since Guericke used two different Latin words to describe the large and small drops (*guttulis* and *bullulae*).

Halley was of the view that water vapour rose because its 'atoms' had expanded into bubbles, but he later stated unambiguously in a paper in 1688 that clouds are composed of *drops*, even visible drops.

Leibniz argued that if the air in such bubbles was less dense than the air surrounding it outside, the bubbles would be compressed until the densities were the same. He then suggested that some heating process like fermentation warmed the bubbles and that they did not cool because 'the air outside will strike it and impart its heat to the bubbles'. He then gave some analogy about moving one's hand through warm or cold water. It was all rather vague.

The rector of Upminster, William Derham, gave the Boyle Lectures in 1711 and 1712, in which he enthused about vesicles so much that some thought they were his idea. Indeed he said he had actually observed them.

Newton called cloud particles 'globules', which could mean drops or bubbles.

John Rowning looked sceptically at the bubble theory and asked why the air inside should be lighter than that outside, and if it was initially lighter why should it stay so, rightly doubting Leibniz's explanations. He also asked why the bubbles did not get bigger as they rose into less dense air (and burst).

Desaguliers equally doubted the concept of bubbles but proposed that the drops/vapour rose due to static electricity (see later).

Christian Gottlieb Kratzenstein also constructed a cloud chamber and, when it was illuminated by sunlight, he noticed a flash of iridescent colours as the drops formed. He took these to be the colours seen on soap bubbles, previously investigated by Newton in his light interference experiments with thin films. In fact the colours were due to refraction, not interference. Using Newton's measurements of soap bubbles, he did calculations assuming bubbles, but these showed that the vesicles would be too heavy to float.

De Saussure was certain that clouds were vesicular, forming this view after looking with a lens at cloud particles over the surface of a cup of dark coffee, their behaviour convincing him they were bubbles. He also looked at cloud particles with a lens while on a mountain, using a black surface to make them more visible, and concluded the same. He repeated Kratzenstein's cloud-chamber experiment and decided that the colours he saw were because the cloud vesicles were like soap bubbles. He dismissed the calculations that Kratzenstein had made, saying that the lightness of the bubbles was due to 'a very light adhering atmosphere' and that 'most physicists believe that almost all bodies are surrounded by a fluid, much rarer than air, that adheres to them'. He went on to point out that Priestly had discovered 'airs' of many kinds, supporting this view. But later de Saussure came to favour the idea of static electricity as the cause of the particles rising, reflecting that outbreaks of heavy rain occur during lightning because the vesicles are suddenly deprived of (electrical) support and fall as rain.

But de Saussure, unlike most others, thought to ask the question of how the vesicles might be formed. He did not know, but he wrote about how he would stand in the mountains and watch as small wisps of cloud suddenly formed over the meadows below where a second before there had been nothing. He reflected that it was curious to think that in air, saturated with transparent vapour, there was lacking just one condition for the vapour to change into vesicles the instant the correct condition existed and form a cloud. This did not answer his question, but it does remind us how puzzling it must have been to see a cloud form out of, apparently, 'thin air'. It is little wonder the Greeks thought air was turning into water, and that de Saussure was intrigued but unable to explain the phenomenon. The concept of dew point and cooling by ascent was still a long way off.

Are clouds composed of droplets?

The belief in vesicles was not universal, however, and Georg Wolfgang Krafft (1745) wrote giving a different explanation for why clouds remain suspended. He said that, once in the air, such very small particles, even if they are heavier than air, can remain suspended there because they were infinitely small and because of the viscosity of the air. He also suggested that the air was constantly in motion due to the action of the sun.

A little later, using a drawing to illustrate the path of a falling particle, Franklin also introduced the matter of viscosity and discussed the forces acting on very small droplets as they descend. It got little attention.

Apart from this brief, unnoticed detour into viscosity, the idea of vesicles as the cause of cloud particles remaining aloft continued. Nothing new was then said until the mid nineteenth century, when the optics of drops was investigated, in particular in the fogbows sometimes seen opposite the sun in cloud or fog. Auguste Bravais at the Ecole Polytechnique believed in the vesicle theory and in 1845 did some elaborate geometrical calculations to show how the bows were formed by hollow droplets. But ten years previously the Astronomer Royal, Sir George Airy, had shown that the geometrical approach was inadequate, the fogbow being produced by diffraction (light being bent round the drops rather than being reflected and refracted through the raindrops as in a rainbow). There is usually little colour in fogbows, but if there is any it is due to interference, not to the thin-film effects that cause colours in bubbles. So yet another line of enquiry failed to demonstrate the existence of vesicles.

Little experimental work had been done to explore the matter of drops versus bubbles but Augustus Waller now made some observations by collecting cloud droplets on films of oil and on spiders' webs and reported that he could see external objects reflected through the drops, showing that the droplets were complete spheres. He also noted that he did not see the drops 'bursting' on contact, as de Saussure had claimed he had. As regards the cloud-chamber tests of Kratzenstein, Waller correctly pointed out that the colours seen were due to diffraction, not to thin films.

Perhaps the biggest problem of all was the lack of any consideration as to how vesicles might possibly form in the first place. De Saussure had made the vague suggestion of 'crystallisation', but what did that mean, and there were no other suggestions that were in any way sensible. There were, however, a number of attempts to show that vesicles could *not* form by any means whatsoever and that even if they did they would not last for long because of surface tension.

Cloud droplets are not suspended, they fall

In his first book, Dalton (1793) described how he believed the process of cloud formation worked, acknowledging that the idea had been suggested to him by John Gough of Kendal in the English Lake District. (Gough was blind and there were references to him by the poets Wordsworth and Coleridge.) Paraphrasing Dalton's words (for brevity), he suggested that when vapour condenses ('precipitates' in his words), many exceedingly small drops form a cloud. These are 800 times as dense as air and descend very slowly due to the resistance of the air. If the drops fall far enough and into air 'capable of imbibing vapour' they 're-dissolve' and the cloud disappears. This suggests that Dalton (or Gough) understood there was a level at which condensation occurred, although the significance of this was not grasped at the time.

Luke Howard (1803) used this concept in his first presentation on clouds, supposing that water vapour rises and is condensed into droplets that then begin to fall until they reach warmer air and evaporate, using this to explain the flat bases of cumulus clouds. He went on in a less precise vein to say that he thought cumulus clouds were bulbous in shape because the droplets were electrified and mutually repelled each other, and that other forms of cloud were explained by different patterns of electric field. He persisted with this erroneous view all his life.

Recognition that clouds are composed of droplets that fall all the time, albeit slowly, was an important step, the precursor of which had been Kratzenstein's cloud-chamber observations. As to how fast they fell, there was little said and only rough, inconclusive calculations were made. But there was also a dawning appreciation that the air in which the drops were falling might also be moving upwards, due to convection.

It had to await the American, James Pollard Espy, who in 1850 put to rest the worry about the 'suspension' of clouds by saying, 'It is not necessary to inquire, as is frequently done, by what powers are clouds suspended in the air, unless it can be shown that they are suspended, of which I think there is no probability. We have every reason to believe that the particles of cloud begin to fall through the air as soon as they are separated from the up-moving column of air by means of which the cloud formed.' Further progress had to await the development of cloud microphysics in the twentieth century (see next chapter).

So by this point it was known, roughly, that clouds were composed of small droplets and that they slowly descended. But under what conditions did clouds form?

What causes clouds to form?

In the Renaissance period, thoughts on how clouds formed, along with how rain-drops formed within the clouds, tended to be bundled together as one question, but it is possible to disentangle the emerging ideas about the two different processes. We will look first at what ideas were evolving about the conditions under which clouds formed, then in the next section we can go on to see what the thinking was about how rain forms within the clouds.

The role played by mountains

Through observation it would have been noticed that mountainous places often had more rainfall and snowfall than the lowlands. This is now known as *orographic* precipitation and we know it its caused by the cooling of the air as the mountains oblige the wind to ascend in order to pass over them. Although it was not yet understood that ascending air cools due to expansion and condensation can result, nevertheless Halley (1693) wrote an account of the hydrological cycle that includes a reference to 'mountains, on the tops of which the air is so cold and rarefied as to retain but a small part of those vapours as shall be brought thither by the winds'.

The most useful observations came from Du Carla (Ducarla-Bonifas), who listed many examples of winds that produce rainfall on the windward side of mountains. In particular he reported that the Cordillera de los Andes removed most of the moisture from the easterly winds, such that by the time the wind got to the western side in Peru it was 'like a dry sponge' drying everything in its path. In 1781 he proposed a thought experiment in which he supposed a wall to be built several thousands of metres high on the meridian of the Observatory of Paris. With a constant easterly wind blowing, he said, on one side would be 'a perpetual deluge' while on the other side, not far to the west, there would be 'an absolute desert'. If the wind reversed and blew from the west the climates would be reversed. He went on to say that if a wall, or a mountain range, of lesser height was present, it would only partly dry the air. Even a slight elevation will cause some precipitation of the moisture from saturated air. Generally, Du Carla recognised that ascending winds are wet, those descending are dry.

Clouds formed by convection

The first suggestion that convection is an important process in cloud formation comes in a letter from Franklin to John Mitchell in 1749, in which he shows awareness that differential warming of the land, and thus of the air in contact with it, can occur, and that the warmed air rises, the colder, denser, air above descending. This was restated in 1765 by Johann Heinrich Lambert (better known for photometry), but in connection with the effects of atmospheric heating on the measurement of heights using a barometer, although he made no reference to this regarding meteorology and his paper was probably seen by only a few people. Then Du Carla (1780) expressed the view that whenever the ground is warmer than the air, a rising air current is produced and clouds form in such currents. But it was still not appreciated that the rising air *inevitably* becomes cooled by expansion. However, Lambert had written a year earlier describing the upward motion of warmed air due to its smaller weight and that as it rises the 'particles of fire' get further and further apart, their density reducing, thus making the air colder.

De Saussure knew of Du Carla's writings and developed a theory about the rise of vapour through convection, referring to 'columns' of rising air, suggesting that he had perhaps some insight into what we now call convective cells. He also observed that ascending air must be replaced by air coming from a cooler region.

Cooling by rarefaction

Du Carla believed that the rarefaction of ascending air leads to condensation and said in 1781 that air cannot rise without 'becoming colder' and 'more rare' and without losing 'on both counts' its ability to hold substances in solution. But he did not appreciate that it was the 'becoming colder' alone that caused the condensation; he thought *both* caused it. These ideas got little attention. Du Carla, according to Middleton (1965), was also the first to suggest that air can be lifted by cold air pushing in beneath it (foresight of the cold front), but he only applied it to the one special case in Africa where the sun warms the rarefied air over the 'Mountains of the Moon' in April each year which is raised by the colder, winter, air from the north. (Mountains of the Moon is the mythical name of mountains in what is now Uganda, Tanzania and Kenya.) But here the ideas stopped. So often, it seems, good ideas are proposed and then ignored or forgotten for a hundred years or more.

In 1755, William Cullen at Glasgow University observed that air which was quickly expanded in the lab cooled, but wrongly attributed the cooling to the evaporation of water in the apparatus. It had to wait until 1779 for Lambert to explain the phenomenon and to link it also to the heating of air by compression. But even then it was explained in terms of the relative concentration of 'particles of fire' – the elastic fluid that heat was imagined to be composed of. However, Lambert went further and linked these facts to the cooling of air as it rises. Starting from the current understanding of thermals, as demonstrated by Du Carla and others, Lambert explained that such an upward movement of heat goes on all the time, the heat from the sun rising, keeping the atmosphere warm throughout its depth. The upward motion of the thermal 'depends on its smaller weight compared with its surroundings, the "particles of fire" getting farther and farther apart as they rise, their density decreasing making it colder'. But again no meteorologists noticed this.

Industrial pointers to cooling by expansion

Next to approach the subject, from a different angle, was Erasmus Darwin, Charles Darwin's grandfather. In a paper with the long title 'Frigorific experiments on the mechanical expansion of air, explaining the cause of the great degree of cold on the summits of high mountains, the sudden condensation of aerial vapour, and of the perpetual mutability of atmospheric heat' (Darwin 1788), he describes observations that demonstrate this:

(1) The air from the receiver of an air gun cools when the gun is fired.
(2) The air in a vessel that is being exhausted by an air pump cools.
(3) Ice collects on the outlet of a 'hero's fountain' in Hungary with a head of 260 feet (79 m).
(4) There is also a more obscure example involving 'the air vessel of a water-works' which smoothes out pulsations from a reciprocating pump.

From these observations, Darwin concluded that whenever air is mechanically expanded 'it becomes capable of attracting heat from other bodies in contact with it', and he said that this explained the coldness of mountains and the upper air. This was, at the same time, confirmed through balloon ascents. Darwin also noted that when condensation occurred in the atmosphere it became warmer. This was the *latent heat*, discovered by Joseph Black, not formally published, but known to those of the Lunar Society (Darwin, Watt, Priestly, Deluc and others) which met monthly in Birmingham when the moon was full (to ease travel at night). The part played by the latent heat of condensation in weather phenomena was not, however, understood until the nineteenth century, into which we now move.

Cooling through ascent

Although good ideas had been accumulating over the past hundred years, such as the realisation that the expansion of air causes cooling, by the year 1800 these hard-gained advances had been in large part forgotten and were being replaced by the bad ideas of the past, such as the pseudo-chemistry of Deluc. But twenty years into the century, the idea of ascending air currents returned, reintroduced by Leopold von Buch in Germany, who said that 'the principle of the ascending current of air should really be called the key to the whole science of meteorology'. Angelo Bellani quoted Humphry Davy on the subject in support of the view, but the idea was confused in many people's minds. As Middleton says, the phenomenon of cooling due to expansion can only be understood fully by doing the mathematics of the process and all those working at the time were 'not of that type of mind'. (Such an understanding requires knowledge of Boyle's and Charles's laws and of the kinetic theory of gases. These are explained in the next chapter.)

However, in 1823 Siméon Denis Poisson derived equations relating temperature and pressure of a gas compressed or expanded adiabatically (that is, without any exchange of heat with its surroundings). This work only applied to unsaturated air, for it was soon realised that saturated air behaved differently because condensation occurs which releases latent heat and this of course changes the balance.

A little later, James Pollard Espy (1835) published his thoughts (in an obscure journal) on the ascent of humid air and the resultant growth of clouds. With simple equipment, he had found that dry air cooled about twice as much as saturated air for a similar change of pressure. From this he concluded that when water vapour in an ascending current of air begins to condense into cloud there is a release of heat that raises the temperature of the air which also causes it to expand as far as the cloud extends. This in turn prevents the ascending air from becoming cold as quickly as it would if it was dry air and so allows it to continue its ascent. Further, if the ascending column continues to be fed by warm moist air from the surface, equilibrium will not be restored.

Then Elias Loomis (1841) made an inventory of the processes by which clouds can form, including:

radiation (dew)
warm air coming in contact with a cold land or sea surface (fog)
mixing of warm and cold currents (light rain)
the sudden lifting of air into a higher position, divided into flow up a mountain, volcanic
    activity, whirlwind (cyclones)
when a hot and a cold current moving in opposite directions meet, the colder air displaces
    the warmer, which is suddenly lifted from the surface and thereby cooled, part of its
    vapour being precipitated

Only the last, did he think (correctly), produced significant rain in mid latitudes, and people now turned their attention to this last item – the meeting of warm and cold masses of air – or what we would call depressions. The American engineer William Redfield established their circulatory (or 'giant whirlwind') motion, with proof from William Reid, Governor of Bermuda, who had access to Admiralty records.

These ideas were then improved on further by Loomis. At this time there were still few observations of barometric pressure, temperature, wind direction and rainfall over a large area, but Loomis collected all those available from the USA and analysed a particularly large storm that occurred on 20–23 December 1836. By studying the wind patterns he concluded that the northwest (cold) wind displaced the southeast (warm) wind by flowing under it, the southeast air mass 'escaping' by ascending from the surface. He produced a cross-sectional drawing that shows what we now call a frontal surface (see Figs. 4.12 and 4.13 in Chapter 4).

But one matter remained to be investigated and explained.

The need for nuclei

Once it became known, through the Guericke cloud-chamber experiments, that the expansion of air caused cooling and that the cooling produced a cloud, it had been assumed that the cloud formed simply because the air had reached dew point; nothing else seemed to be involved.

But then a century after the Guericke discovery, while P. J. Coulier was doing similar experiments in France in 1875, he came across an anomaly. Using a closed system, consisting of a glass flask containing water connected by a tube to a hollow rubber ball that could be alternately expanded and contracted, he found that after a few cycles of expansion and contraction, fog stopped forming. He also got the same result if the flask was left undisturbed for a few days. The same thing happened if he shook the flask for several minutes before contracting the air: no cloud would form. This went against expectations. If this 'inactivated' air was replaced by fresh air from the room, clouds would again form. If the fresh air was first filtered, no clouds would form. Coulier deduced that something additional in the air was necessary for drops to form and assumed that it must be dust of some form, since it could be filtered out. As happened too often in the science of the time, this result was forgotten and had to be rediscovered by John Aitken (1880), who demonstrated similar results to the Royal Society of Edinburgh, believing it was original work. When a letter appeared in *Nature* he realised that he was not the first to have discovered this and freely acknowledged so. However, Aitken had taken the tests further, an important find being that the fewer the 'dust' particles, the larger the drop sizes and the fewer their number, suggesting that the available water vapour was shared amongst the particles.

Aitken then investigated what the dust might be composed of and concluded that most probably dried sea spray in the form of fine dry powder was the main source of nuclei, other possible causes being volcanic and meteoric dust, along with smoke from combustion. He stressed that it is not the dust motes visible to the eye floating in a sunlit room, but much finer quite invisible particles, ever-present in large numbers but never observed. Tests on a variety of finely divided substances showed that they differed greatly in their nucleating power, hygroscopic nuclei being the most active, which includes both salt and the hygroscopic products of combustion.

Aitken (1889–90) then summarised his conclusions as follows:

Water is deposited on many nuclei even in unsaturated air, producing tiny droplets that are seen as haze.

As the amount of water vapour increases and the dew point is approached, the droplets grow and the haze becomes thicker.

At saturation a change occurs, the more abundant vapour now condensing onto the larger drops, in preference to the smaller, and a cloud forms rapidly with different properties to haze.

However, while there is an initial tendency for the larger drops to grow at the expense of the smaller, this is offset by the fact that as the larger drops grow the concentration of the hygroscopic substance (in the drops) decreases, while if the smaller drops evaporate, the concentration increases, acting as a moderating or negative-feedback influence.

If the air is slightly supersaturated (101% relative humidity), even non-hygroscopic nuclei can form drops. However, since the amount of supersaturation required to form droplets on non-hygroscopic nuclei increases greatly as the drops get smaller, and as supersaturation in the free atmosphere is always very small, non-hygroscopic nuclei will never be very active. Nor will gaseous ions or aggregates of water molecules.

But what turns the cloud droplets into precipitation?

Emerging ideas about precipitation

It was now known how cloud droplets form. But what turns them into raindrops or snow or hail? To follow the path of discovery we have to go back to the time when it was thought that cloud particles were small bubbles.

The growth of raindrops

The compression theory

Descartes, along with Desaguliers and Deluc, initially pondered on whether the wind, by compressing the clouds or pushing them together, might make the drops adhere to each other and thereby grow in size. Deluc, around 1786, suggested that the 'vesicles' came into contact and joined, eventually 'bursting' like a large soap bubble producing drops which then go on to collect more vesicles as they fall. The idea that clouds were compressed, forcing the particles together, was common throughout the period. Desaguliers suggested that lightning might do this and he also thought that when a cloud 'collided' with a hill the particles were forced together. The Reverend Hugh Hamilton (Dublin) and Deluc believed that clouds were compressed by a chain of mountains, through the action of a contrary wind, or compressed simply by the resistance of the clouds to the wind. By 1786 Deluc was less certain about the compression theory, believing instead that rain clouds differ from non-rain clouds because 'so much vapour is being formed that the vesicles come into contact and burst like soap bubbles'. Pardies also believed that clouds were composed of little bubbles that simply burst to make raindrops.

The contrary view was put by Halley, who suggested that the winds might rarefy the air when they blew from one place in opposite directions and, being rarefied, the atoms of air are fewer and so it is easier for the droplets to combine. Halley (1693) also wrote that he thought 'saline or angular particles' may aid in the more speedy condensation into rain, while 20 years later Edward Barlow made the suggestion that the production of rain appears to require 'essences of minerals and sulphurous exhalations out of the bowels of the earth', a view noted also by Musschenbroek. With hindsight these suggestions could be seen as precursors of the idea of nuclei required to form cloud droplets, but it seems unlikely and in any case they were discussing the formation of raindrops from cloud, not cloud droplets from vapour.

Increase of rainfall near the ground

While Aristotle had the view that the largest raindrops and hail originate nearest to the ground, no one in the seventeenth century thought this, the general opinion being that the drops increased in size the further they fell. Two processes were thought to cause the growth of drops, the first being the coalescence of the drops as they fall, the second being condensation onto the falling drops. However, to confuse matters, some experiments were done (between July 1766 and June 1767) by the physician William Heberden (the Elder) in which he exposed three identical raingauges at different heights nearby to each other and found that the higher one (on the roof of Westminster Abbey) caught 46.5% less than one in his garden while the one on the roof of his house caught 19.8% less than the one on the ground. This led Heberden to speculate that some unknown electrical effect was involved, although those who knew more about electricity were less convinced, including Franklin. Shortly after the Heberden experiment, Daines Barrington operated two raingauges in north Wales, one on a hillside 1350 feet (411 m) higher than the other. The difference in catch was much less than Heberden's, making Heberden conclude that the difference does not seem to depend on the depth of atmosphere through which the rain falls. There had been speculation that the experiment in London showed that the drops might grow considerably in the last few hundred feet of their fall, perhaps by condensation onto the cold drops. This debate continued until eventually in 1861 Jevons demonstrated unambiguously that the difference in catch was entirely due to wind effects. (This matter is dealt with fully in Chapters 7 and 8, on rainfall measurement, with appropriate references.) So if raindrops do not grow significantly in the last part of their fall, how do they form in the clouds?

Coalescence or condensation?

Two possible means of raindrop growth were visualised: the cloud droplets can grow either by combining with each other or by water vapour condensing onto them – or both. An important step was taken in 1870 by Lord Kelvin (William Thomson) who showed that, because of surface tension, the vapour pressure over the spherical surface of a drop of water is greater than it is over a flat surface, by an amount proportional to the reciprocal of its radius. Not all droplets are of the same size and the variation in vapour pressure with drop size results in evaporation from the smaller drops and condensation onto the larger ones. But in 1877 Osborne Reynolds concluded that condensation would proceed too slowly to account for the size of drops observed in reality. He concluded that the second process – coalescence – accounted for most of the growth and that this continued until the drops grew to such a size that they were broken up by the resistance they encountered during their fall through the air.

Moving now just into the twentieth century, Philipp Lenard investigated these questions and in 1904 wrote that drops can be grouped into three sizes:

small drops, whose falling speed is proportional to their radius squared
medium-sized drops, up to 1.1 mm, the falling speed of which is proportional to the
    square root of the radius (because of the effects of turbulence)
drops large enough to be deformed by the effects of the passage of air past them

He reported that he had observed deformation using flash illumination in 1887. In fact Gustav Magnus had made similar observations back in 1859, but Lenard made complex equipment which allowed him to measure the size of drops suspended in a vertical upward air current. He found that drops with any diameter over 4.5 mm fell at 8.0 m s^{-1}, this being less than twice the speed of a drop of 1 mm diameter.

Lenard then went on to calculate that if drops vary in size, they should collide often (because they fall at different speeds), but because many clouds do not produce rain, he assumed that many of the collisions did not result in a blending. As to why this was, Lenard hypothesised that it might be due to electrical charges on the drops causing them to repel each other, or 'because of a layer of air absorbed on the drops'.

Julius Wiesner, in 1895, working in the tropics, measured the size spectrum of raindrops by catching the raindrops on filter paper powdered with a dye, measuring the stains that resulted. In 1905, Albert Defant, using the Wiesner method, measured many thousands of raindrops and found that the weight of the drops clustered in multiples of some basic small weight, in particular 1, 2, 3, 4, 5, 6, 8, 12, 16 and so on. The main maxima were at 1, 2, 4 and 8. This was true for all types of rain. This clearly supported the view that drops grow by combining with other drops, and Defant believed that condensation onto larger drops would be too slow and that most of the drop growth was due to coalescence.

These results also suggested that drops of similar size were more likely to combine than unequal-sized drops. In 1908 Wilhelm Schmidt explained this pattern of drop sizes as being caused by the hydro-dynamical attraction produced when two drops are falling side by side at similar speeds, and calculated that the distances and times involved would support this hypothesis.

The work pioneered by these late nineteenth- and early twentieth-century researchers was taken up through the twentieth century, growing into the flourishing field of investigation that we now call 'cloud physics', of which more will be said in Chapter 4.

The growth of hailstones

Hail was the most difficult form of precipitation for early thinkers to understand. The two main questions facing those seeking an explanation of it were: How does the cooling take place (where does the latent heat of fusion go)? and What keeps the hailstones aloft for long enough to freeze?

Chemical explanations

Descartes noted that hailstones often had what looked like snow inside them and suggested that the larger stones grew when the wind drove many snowflakes together, which then partly thawed and refroze. He was not happy with this view but at least the idea was not too fanciful. Others like Gassendi believed that the upper atmosphere was cold because 'nitrous corpuscles' of cold are plentiful up there in the summer but less so in the winter, when only snow was produced. Others, like John Mayow and John Wallis in the mid seventeenth century, believed in the nitre idea because hail occurred with lightning and, at the time, lightning and a gunpowder explosion were seen as somehow related, the thought being that if nitre was in the air producing lightning it probably formed a freezing mixture with snow. Musschenbroek also observed the association of hail with lightning and believed the latter was made of sulphurous matter and 'spirit of nitre' which 'excite a horrible cold'. Supporting these views was the common idea that it needed more than just cold to freeze water, one belief being that it occurs when the water mixes with 'very slender corpuscles' from the atmosphere, resulting in a form of fermentation which 'drives away the heat', causing the water particles to adhere to each other. According to Middleton (1965) this view probably stemmed from observations of supercooling, unknown at the time (see below).

The chemical explanation slowly waned during the eighteenth century and was replaced by a more realistic physical account based on the observation that hail occurs mostly in summer. Deluc's view was that the formation of hail was proof of the great vertical extent of vapour and that the snowy core of a hailstone suggested that it formed at such a height that the vapour it encounters in its fall freezes round it. Both Franklin and Monge believed the same, although Monge thought drops rather than vapour accreted round the snow.

Electrical explanations

A problem with Franklin's kite experiment and the association of hail with thunder was that it inevitably led to the speculation that hail was produced, somehow, electrically. Guyton de Morveau rejected the 'nitre' chemical explanation, asking why none of the suggested 'salts' was found in the hailstones, while promoting instead the idea that electricity was the driving force, but his explanations were hazy. There was then some argument that charged drops would repel each other, so how could it be that an electric charge was involved? Deluc later abandoned his earlier, more reasonable, view and turned to chemical and electrical explanations. Even Volta believed in an electrical explanation, but in his case the electrical charges explained how the hailstones remained suspended long enough to grow, bouncing up and down above the cloud through mutual repulsion until they became too heavy and fell. The cold, he thought, was due to very rapid evaporation and the supposed effects of electricity in promoting evaporation. Much time was wasted speculating in this way. Indeed it is frustrating to read of the development of ideas during this three-century period, which often amounted to little more than conjecture. I will not waste space describing all of the related vague ideas.

Rising air currents

Fortunately by around 1820 a suspicion was growing that the link between hail and lightning might mean no more than that they were both produced by the same conditions – association, not causation. Henceforth ideas were centred on mechanical explanations, in particular rising air currents. Such ideas had in fact existed earlier but had been ignored. For example Du Carla had a theory in 1780 based on rapidly rising columns of air which carried its condensation products to a height where they froze into 'little snowy globules' which grew, possibly remaining suspended for hours due to the strength of the rising air current. Finally, as they fell, slowly at first, they met 'aqueous molecules, coagulated by secretion and cold' which joined the globules, spread out and froze. Du Carla was also able to explain how hail occurred mostly on summer afternoons since it is in this season that the differences of surface temperature are greatest and the rising currents the most intense. Yet Du Carla's writings appear to have been little known outside of Geneva, and we had to wait for the concept of thermals to be rediscovered later.

Supercooling

By the mid nineteenth century, the existence of supercooled cloud droplets began to be recognised. As early as 1783 de Saussure had noted that drops can remain unfrozen well below freezing point, but this was forgotten for another 90 years.

(What are we forgetting today?) When Wähner rediscovered this in 1876 he also recognised that raindrops too can be supercooled, instantly freezing into a glaze frost as soon as they came into contact with a surface. It was but a short step from this to realise that snow pellets falling through supercooled cloud might collect ice around them as they fell. This seemed to overcome the problem about how the latent heat of fusion was disposed of quickly enough.

While Vogel and Karl Nöllner picked up on the idea of supercooling, it was Maille (1853) who published the first sound theory, in which he proposed that snow pellets formed high in the atmosphere and then descended through supercooled cloud. He went on to add that in the case of large hailstones there can be alternate layers of snow separated by denser layers of ice, and visualised a process in which the stones moved up and down under the competing forces of gravity and a strong current of uprising air due to thermals.

Support for these ideas came from a balloon ascent in 1850 by Barral to nearly 19 000 feet (5790 m), during which he rose through a layer of cloud from 11 250 to 18 990 feet (3430–5788 m) in which the temperature fell from $-0.5\,^{\circ}C$ to $-10.5\,^{\circ}C$, the lower cloud being supercooled, and the upper part ice crystals.

But the question of the latent heat of fusion of ice was not solved by the discovery of supercooling because in the atmosphere supercooling never exceeds a degree or so, meaning that most of the water would not freeze when it contacted a snow pellet; some other explanation was needed. Before such an explanation was forthcoming, Osborne Reynolds suggested that large hailstones form through coalescence of smaller ones. It was not until the mid twentieth century, however, that it was understood that the loss of heat was achieved through a combination of simple conduction into the cold air above the freezing level combined with evaporation (see Chapter 5).

Dew and frost

Moonlight nights are clear nights, which led the ancients to the conclusion that the moon was the cause of the dew. I wrote earlier of how Aristotle was seen as not being a good scientist because he failed to make observations or do experiments. But in the case of dew and frost he got closer to the truth than anyone else for the next two millennia when he said that any moisture evaporated during the day, that does not rise far, falls again when it is chilled during the night and is called dew. Aristotle also knew that dew and hoar frost form in clear, calm weather. Aristotle did not of course use the word 'evaporation' and knew nothing about water vapour or what air was, but he did make a good guess and did look out of the window. For someone so criticised for his lack of observation this is at least a small reprieve.

Does dew rise or fall?

No one improved on Aristotle's idea for 1900 years until Antoine Mizauld added some well-argued modifications, including the observation that dew does not form under a tent, and more importantly he compared the formation of dew with the process of distillation. So by the seventeenth century it was understood that dew was condensed vapour, but two questions remained. First, did the vapour come from the earth or from the air, that is did the dew rise or did it fall (as droplets)? Second, what caused the cooling – did the dew cause it, or did cooling cause the dew? Such understanding was hampered by an incomplete knowledge of what water vapour was. In an attempt to answer these questions some good experiments were done, one of the more important by Cisternay du Fay, who also exchanged letters with Musschenbroek on the matter. Both exposed bowls, some coloured, some white, some left bare, and found that the darker bowls collected most dew. They also exposed glass, porcelain and polished metal, the latter collecting very little dew. But they were unable to explain their findings because at that time the idea of invisible radiation from all bodies above absolute zero in the infrared band was unknown. Because they found that dew formed on the underside of plates suspended in the air as well as on top, Du Fay asked how this occurred if dew 'fell'. He went on to liken this to the way moisture forms on a cold bottle of wine brought from a cellar into a warm moist room, but used this only as a simile, not appreciating that the processes at work were one and the same (condensation onto a cold object).

Readers will remember the experimental work of Charles Le Roy of Montpellier with his glasses of cooled water (described earlier) that acted as his dew point detector. From these experiments, Le Roy concluded that the 'dew' that formed on the glasses came from the air. Based on these experiments he concluded that dew always formed when the dew point temperature is reached. He also noticed that dew formed less in town centres than in the surrounding country and concluded that this was because it stayed warmer there.

There were mild distractions, such as that by Wolf who in 1762 drew attention to the similarity between frost patterns on windows and salt crystals that form in evaporation pans. This brought back the old idea that freezing requires some 'frigorific particles' of some (unspecified) salt.

Then, between 1767 and 1780, Patrick Wilson, working in Scotland, exposed identical thermometers at night, one on the snow, another suspended 30 inches (76 cm) above it in the air, and found that the temperature indicated by the thermometer on the snow read lower than that in the air. Wilson wondered if this was because of cooling by evaporation at the snow surface, one of the current ideas being that cooling by evaporation resulted in the formation of dew. (This is odd thinking because they were saying in effect that evaporation produced cooling and that this produced condensation.) But he proved that this was not the case by fanning the thermometer on the snow with a sheet of brown paper on a stick, whereupon its temperature rose to near that of the air. Wilson, however, was unable to explain this. Then in 1788, James Six, inventor of the Six's maximum and minimum thermometer, used his new creation to measure a temperature profile up to 220 feet (67 m), using the tower of Canterbury Cathedral, and found that it was usually colder near the ground at night, warmer during the day, and that this was most pronounced in cloudless conditions. Despite what Wilson and others had said, Six concluded that it was colder near the surface because of the coolness the dew acquired in its descent, and he also wrongly concluded that evaporative cooling contributed to the effect. But these views again raised the question as to whether dew causes the cold or vice versa, the view at the time being that the cold was caused by the dew. Patrick Wilson also continued his experiments, but they too only added to the confusion, lacking the one key fact of radiation loss at night.

Cooling by infrared radiation

Everyone was confused when, in 1774, Marc-Auguste Pictet, working in Geneva, showed that IR radiant heat was reflected by mirrors, including parabolic mirrors. He set two parabaloids facing each other, a flask of snow at the focus of one and a thermometer bulb at the focus of the other. The thermometer cooled, and Pictet wrongly concluded that not only could heat be radiated, but that cold could also. Saving the day, Pierre Prevost, a colleague of Pictet in Geneva, produced the theory of the equilibrium of radiant heat, explaining that heat is exchanged all the time between all bodies. It was not, he explained, that cold was radiated to the thermometer from the flask of snow, but that the thermometer radiated heat to the snow. Despite this explanation, Pictet, having made more observation of temperature profiles, still explained the night-time inversion of the temperature gradient as being due to 'evaporation from the soil'.

There was now great uncertainty over the whole matter, but a solution to the problem came quite soon when Benjamin Thompson (Count Rumford) wrote his classic paper (1804) on IR radiation from bodies with various surface properties. Through experiment he had found that polished metal objects radiate much less heat than most other substances at the same temperature. Conversely he found that glass was a good radiator (we also now know that snow is a good radiator). Despite his own tests and their obvious logic, he persisted in the belief that cold bodies radiated cold, as if the inverse process was occurring.

While the idea was accepted that some bodies cool faster than others, the suggestion that it was through invisible radiation was not. John Leslie for example maintained that air never actually contacts any surface but comes nearer to glass than it does to polished metal, claiming that air is always 1/500 of an inch

(50 μm) from the metal. (We can of course argue that he was right in the sense that there is in reality a *laminar boundary layer* about 1 mm thick around all objects although, nevertheless, the air is still in intimate contact with objects within it. The laminar boundary layer is just immune to turbulence and all transfer through it is by diffusion alone. See Chapter 3 and Fig. 3.4)

William Charles Wells, an American doctor living in London, investigated dew in considerable detail and published his findings in his *Essay on Dew* (1814), in which he reported the result of many experiments, some simply repeats of earlier work by others. An important original piece of work was to expose a number of balls of wool weighing 10 grains (0.65 g) on a grass plot, with one of them surrounded by a ceramic cylinder. Those fully exposed collected 1.0 g of dew, that in the cylinder only 0.13 g. Wells concluded that

Whatever diminishes the view of the sky, as seen from the exposed bodies, occasions the quantity of dew whatever it is formed upon to be less than would have occurred if the exposure to the sky had been complete.

He added that this also showed that dew did not 'fall'. He also confirmed that on cloudy nights the temperature of the grass was little or no colder than the air. Wells then considered whether dew 'rises' but concluded that dew appears earlier on objects nearer the ground because the air nearer the ground is colder and because the air is less disturbed.

The idea of nocturnal radiation was difficult to accept for many, but Wells had shown without doubt that

the cooling of the Earth's surface, and of the bodies that accumulate dew, is the result of radiation to space. This radiation is always going on, but can be largely interrupted by clouds: and in the daytime it is overbalanced by the radiation to the Earth from the Sun.

The work of Wells was built on by James Glaisher (1847), who produced a long paper on nocturnal radiation in which he reported measurements of temperature and RH from which he concluded that the formation of dew depended solely on the temperature of the bodies upon which it was deposited, and it never appeared upon them until their temperatures had descended below that of the dew point in their locality, as found by observations of a dry and wet bulb thermometer placed in their vicinity.

At the same time the Italian physicist Macedonio Melloni investigated the matter of radiation by putting thermometers in bright metal boxes and in boxes covered in soot and with them demonstrated a large difference in emissivity. But his most decisive experiment (too complex to go into here, involving an ingenious construction of a large disc shaded by a smaller disc above) demonstrated to Melloni that dew is 'a pure consequence of nocturnal radiation'.

A few, like John Aitken, were brave enough to continue to argue, based on their own experiments, that *some* of the dew came from the ground. But by early in the twentieth century it was generally agreed that dew was simply water vapour condensing from the air onto any object that cooled, through IR radiation loss, to below the dew point of the air, and that cloud cover reduced this loss and inhibited dew formation.

Summary

So by the end of the three-hundred-year period covered by this chapter, all of the basic ideas concerning the production of precipitation had been established:

how evaporation occurs

what water vapour is

that water vapour is condensed into cloud when humid air is cooled to dew point, usually by expansion whenever air is raised

how condensation leads to latent heat being returned to the cloud, thereby warming it, assisting its continued ascent

how tiny cloud droplets form on hygroscopic nuclei and how these then combine to form raindrops

the formation of hail in part through supercooled cloud droplets accreting onto snow pellets (but not fully understood until the twentieth century)

that dew was water vapour condensing from the air onto any object that cooled, through infrared radiation loss, to below the dew point

We move now into the twentieth century and the continued exploration of these processes.

References

Aitken, J. (1880). On dust, fogs, and clouds. *Transactions of the Royal Society of Edinburgh*, **30**, 337–368.

Aitken, J. (1889–90). *Proceedings of the Royal Society of Edinburgh*, **17**, 193–254.

Bacon, F. (1620). *Novum organum* (or *True Directions Concerning the Interpretation of Nature*).

Bernoulli, D. (1738). *Hydrodynamica, sive de viribus & motibus fluidorum commentarii.* Strasbourg.

Bouillet, J. (1742). Sur l'évaporation des liquides. *Histoire de l'Académie Royale des Sciences*, 18–21.

Dalton, J. (1793). *Meteorological Observations and Essays.* London: W. Richardson.

Dalton, J. (1802). Experimental essays on the constitution of mixed gases; on the force of steam or vapour from water and other liquids in different temperatures, both in a Torricellian vacuum and in air; on evaporation; and on the expansion of gases by heat. *Memoirs of the Manchester Literary & Philosophical Society*, **5**, 535–602.

Darwin, E. (1788). Frigorific experiments on the mechanical expansion of air, explaining the cause of the great degree of cold on the summits of high mountains, the sudden condensation of aerial vapour, and of the perpetual mutability of atmospheric heat. *Philosophical Transactions of the Royal Society of London*, **78**, 43–52.

Desaguliers, J. T. and Beighton, H. (1729). *Philosophical Transactions of the Royal Society of London*, **36**.

Descartes, R. (1637). *Discours de la méthode* ... [*Discourse on the Method of Rightly Conducting the Reason, and Seeking the Truth in the Sciences*]. Leiden.

Du Carla, M. (1780). *Histoire naturelle du monde*. Genève: Du Villard Fils & Nouffer.

Espy, J. P. (1835). Theory of rain, hail, snow and the water spout, deduced from the latent caloric of vapour and the specific caloric of the atmospheric air. *Transactions of the Geological Society of Pennsylvania*, **1**, 342–346.

Franklin, B. (1765). *Philosophical Transactions of the Royal Society of London*, **55**, 182–192.

Glaisher, J. (1847). On the amount of the radiation of heat, at night, from the earth, and from various bodies placed on or near the surface of the earth. *Philosophical Transactions of the Royal Society of London*, **137**, 119–216.

Halley, E. (1693). *Philosophical Transactions of the Royal Society of London*, **17**, 473.

Hamilton, H. (1765). *Philosophical Transactions of the Royal Society of London*, **55**, 146–181.

Howard, L. (1803). On the modifications of clouds. *Philosophical Magazine*, **16**, 97–107.

Krafft, G. W. (1745). *Cogitationes in experimenta et sententias de vaporum et halituum generatione ac elevatione*. Tübingen.

Le Roy, C. (1751). Mémoire sur l'élévation et la suspension de l'eau dans l'air, et sur la rosée. *Mémoires de l'Académie Royale des Sciences, Paris*, pp. 481–518.

Locke, J. (1690). *An Essay Concerning Human Understanding*. London.

Loomis, E. (1841). On the storm which was experienced throughout the United States about the 20th of December, 1836. *Transactions of the American Philosophical Society*, **7**, 125–163.

Maille, P. H. (1853). Nouvelle théorie des hydrometeors. Académie Royale des Sciences. *Séance Publique du lundi 8 décembre 1834. Annonce des prix décernés par l'Académie des Sciences pour l'année* (1834), 5–7.

Middleton, W. E. K. (1965). *A History of the Theories of Rain and Other Forms of Precipitation*. London: Oldbourne.

Middleton, W. E. K. (1994). *The History of the Barometer*. Trowbridge: Baros Books.

Molyneux, W. (1686). A discourse on the problem; why bodies dissolved in menstrual specifically lighter than themselves swim therein. *Philosophical Transactions of the Royal Society of London*, **16**.

Musschenbroek, P. van (1736). *Beginselen der Natuurkunde*. Leiden.

Newton, I. (1704). *Opticks* (or *A Treatise of the Reflections, Refractions, Inflections and Colours of Light*). London: the Royal Society.

Saussure, H. B. de (1783). *Essais sur l'hygrométrie*. Neuchâtel, pp. 257–258.

Thompson, B. (1804). An enquiry concerning the nature of heat and the mode of its communication. *Philosophical Transactions of the Royal Society of London*, **94**, 77–182.

Wallerius, N. (1738). De ascensu vaporum in vacuo. *Acta Literaria* (Uppsala), **4**, 339–346.

Wells, W. C. (1814). *An Essay on Dew and Several Appearances Connected With It*. London: Taylor and Hessey.

Part 2

Present theories of precipitation

For most of human history, as described in Chapter 1, hard facts were scarce and speculation widespread. By the eighteenth century, however, sounder ideas were at last slowly forming (Chapter 2). And now, in the twenty-first century, detailed explanations are plentiful. Chapters 3, 4, 5 and 6 review our present knowledge, Chapter 3 starting with the very basic processes.

3

Basic processes

Evaporation

In the preceding chapter it was shown how the discovery by Robert Brown in 1827 of the continuous vibration of very small particles, now known as *Brownian movement* (or *motion*), led to the kinetic theory of gases. This is central to the process of evaporation. The molecules of liquid water are much closer together than those of a gas and are separated from each other by just slightly more than the diameter of one molecule. In such close proximity, the atomic particles strongly attract each other by Van der Vaals forces (electrical attractions between molecules), but this force reduces rapidly as their separation is increased. In water vapour, the spacing between molecules is ten diameters or more, depending on their concentration (or *vapour pressure*, VP – that fraction of the total pressure due to the water vapour alone – measured in any of the usual units of pressure, such as millibars or hectopascals) and the attractive force is then extremely small. To produce water vapour from liquid water, the distance between the molecules has to be increased, and to achieve this, work has to be done against the binding Van der Vaals attraction.

During evaporation (at any temperature) water molecules 'boil off', due to Brownian movement, at a rate proportional to the absolute temperature. Some, however, by chance, due to their random motion in the air, find their way back to the water surface, the numbers doing so depending on their concentration in the air which is directly related to the VP. Some of the returning molecules bounce off the surface but some re-enter the liquid water. The net rate of evaporation is the difference between the boil-off and re-entry rates. So long as the vapour molecules in the air can move away from the water surface by processes of diffusion (see below) their concentration, immediately adjacent to the water surface, remains low and a continual flow of molecules away from the surface into the air occurs, at a rate set by the absolute temperature.

Basic processes

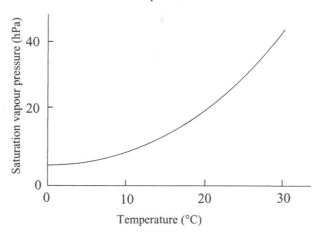

Figure 3.1 If a volume of water is enclosed within a sealed container, the evaporating molecules cannot diffuse away and their numbers, and so the vapour pressure (VP), increase. When the number of water molecules leaving the water surface is the same as the number rejoining the water, the air in the container is said to be *saturated*. At any given temperature this balance occurs at a particular vapour pressure, which is known as the *saturation vapour pressure* (SVP). As the temperature increases so does the SVP, as the graph shows. At 0 °C the SVP is 6.11213 hPa over water.

Saturation

If, however, the water molecules cannot diffuse away, for example if the water is in a sealed container, their numbers, and so the vapour pressure, increase until the two rates are equal. Although the same numbers of molecules continue to escape from the water surface, the same numbers also now return to it and evaporation ceases. The air in the container is then said to be *saturated*. At any given temperature this balance occurs at a particular vapour pressure, which is known as the *saturation vapour pressure* (SVP) (Fig. 3.1). At 0 °C the SVP is 6.11213 hPa over water. There are equivalent SVPs over ice, being 6.11154 hPa at 0 °C. SVPs are available from hygrometric tables.

Units

In addition to VP and SVP, there are a number of other ways of expressing the water vapour content of air. The *relative humidity* (RH) is the ratio (expressed as a percentage) of VP to SVP. Thus 100% RH indicates saturation. This applies just to the temperature of the sample at the time so, without also knowing the air temperature, the RH is not complete information.

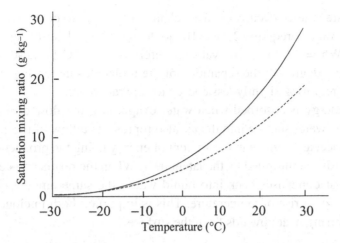

Figure 3.2 Another way of expressing the maximum amount of water vapour that air can contain before condensation occurs is by expressing the amounts in weights, the *saturation mixing ratio* being the largest ratio of water vapour to air at any one temperature, expressed in grams of water per kilogram of air. Not only is this dependent on temperature (as is the VP) but it also depends on the density of the air. The upper curve is for near-surface pressure, while the lower broken curve shows the mixing ratio at around 3000 m altitude. Saturated air at increasing altitude contains increasingly less water vapour.

Other units include:

mixing ratio – the ratio of the *mass* of water vapour to the *mass* of the dry air it is associated with

saturation mixing ratio – the largest *ratio* of water vapour to air at any one temperature (Fig. 3.2)

specific humidity (or *water content*) – the ratio of the *mass* of water to the *mass* of moist air

vapour concentration (or *absolute humidity*) – the ratio of the *mass* of water vapour to the *volume* of the moist air it is associated with, which is in effect the density of the water vapour, hence its additional name of *vapour density*

Cooling moist air (while keeping its pressure constant) increases its RH, and a temperature will be reached when the RH is 100%. Further cooling then causes dew to form. This is known as the *dew-point* temperature. There is a similar temperature, below 0 °C, known as the *frost-point* temperature (or *hoar-frost point*).

Latent heat

The amount of energy required to convert unit mass of liquid water into vapour is known as the *latent heat of vaporisation of water* ('latent' because it is hidden, no temperature change occurring, just a change of state). To convert 1 kg of water (at 10 °C) to vapour requires 2.47×10^6 joules (2.4 MJ) (1 joule per second = 1 watt, or 1 kWh = 3.6 MJ). The value is referred to 10 °C because it varies by about 0.1% per degree C, the separation of the molecules increasing slightly with temperature, requiring slightly less energy to separate them.

Because energy is absorbed when water evaporates, the flow of water vapour away from the water surface into the air also represents a flow of the energy away from the surface, as latent heat, the amount of energy being the product of the mass of water vaporised multiplied by the latent heat. When the reverse process happens and the vapour condenses back into liquid water, the same amount of energy is released, causing a rise of temperature. This is important both in cloud formation and in transferring heat upwards from the surface.

Radiation

In the natural world, evaporation occurs from open water and from wet surfaces such as the ground and vegetation following precipitation, and also from within cavities such as the small pores between soil particles and within the leaves of plants, the water vapour being vented through their stomata. The energy source for these processes is the sun.

Solar radiation

Intensity

Integrated over the whole of its radiation spectrum, the sun emits about 74 million watts of electromagnetic energy per square metre of its surface. At the mean distance of the earth from the sun, the energy received from the sun at the outer limits of the earth's atmosphere, at right angles to the solar beam, is about 1353 W m^{-2} which is known as the *solar constant.*

On cloudless days the diurnal variation of solar radiation is roughly sinusoidal. In the UK in summer (for example) the maximum intensity of the radiation at solar noon is about 900 W m^{-2}. With a day length of 16 hours, the maximum daily energy received is about 9 kWh m^{-2}. However, the quantity of solar energy received is often expressed in megajoules rather than kilowatt hours. Since 1 joule per second = 1 watt, 1 kWh is equivalent to $1000 \times 60 \times 60 = 3.6$ MJ, giving a UK daily maximum insolation of about 32 MJ m^{-2}. This is reduced by cloud cover to a summer average of 15 to 25 MJ m^{-2}, while in winter it varies from 1 to 5 MJ m^{-2}.

Solar spectrum

Figure 3.3 illustrates the spectrum of the sun's emission, both as received just outside the atmosphere and as received at the earth's surface after modification by its passage

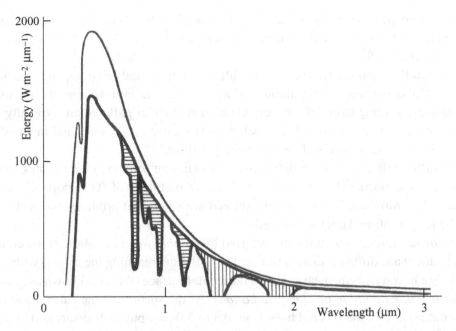

Figure 3.3 The spectrum of the sun's emission just outside the atmosphere is shown by the upper curve. The lower curve shows the reduced spectrum received at the earth's surface after modification by its passage through the atmosphere, radiation in different parts of the spectrum being absorbed, scattered and reflected. The vertical shading shows where absorption occurs in the mid-IR due to water vapour and carbon dioxide while the horizontal shading indicates absorption due to oxygen and water vapour in the near-IR. Absorption also occurs in the UV at wavelengths shorter than 0.4 μm due to oxygen and ozone (not shaded).

through the atmosphere. During its downward journey, radiation in different parts of the spectrum is absorbed, scattered and reflected. For example, some ultraviolet (UV) radiation is absorbed and scattered by ozone and by oxygen while parts of the infrared spectrum are absorbed by water vapour and carbon dioxide. Scattering and diffuse reflection occurs from particles larger than the wavelength of the radiation, such as dust, smoke and aerosols from volcanic activity, while Rayleigh scattering by air molecules and particles smaller than the wavelength gives rise to the blue sky. Clouds can reflect up to 70% of the energy from their upper surface, the lower-altitude clouds (Chapter 4) being responsible for the loss of considerable amounts of solar energy back to space.

The amount of solar radiation reaching the ground at any one time depends on the time of day, season, latitude, cloud cover and atmospheric quality. Taken as an annual radiation budget for the whole planet, of 100 units of incoming radiation just outside the atmosphere, 25% is reflected back to space from clouds and dust, 7% is scattered back to space, 1% is absorbed by clouds, 16% is absorbed directly by the atmosphere (thereby warming it), leaving 51% to arrive at the surface. Ten per cent of this is reflected from the surface (albedo), leaving 41% as available energy at the surface.

In cloudless conditions, the ratio of diffuse to direct radiation depends on the angle of the sun and on the amount of aerosols in the atmosphere, the diffuse component varying from 10% in very clear air to 25% in polluted air; typically it is about 15% (Monteith 1975). Just before sunrise, just after sunset and in cloudy conditions, the radiation is of course entirely diffuse.

To differentiate between radiation from the different sources, various categories have been defined (Meteorological Branch, Department of Transport, Canada 1962). The wavelength limits usually quoted are somewhat arbitrary, particularly at the longer infrared (IR) wavelengths.

Solar radiation, sometimes also referred to as *short-wave* radiation, is the combined direct and diffuse radiation received from the sun reaching the ground without a change in wavelength, measured on a horizontal surface (it is also known as *hemispheric solar radiation* or *global radiation*). At the surface of the earth, most of this radiation is in the spectral band from 0.3 to 3.0 μm, but with absorption bands as in Fig. 3.3.

Terrestrial radiation, sometimes referred to as *long-wave* radiation, is the IR radiation emitted by the ground and by the atmosphere (in particular by water vapour and carbon dioxide – the same gases that also absorb it), without change of wavelength within the (arbitrary) limits of 3 to 100 μm. Included in this is radiation emitted by clouds in the narrow band from 8 to 13 μm, a region in which little radiation is emitted by water vapour and carbon dioxide. These two general categories are subdivided as follows:

  Direct solar radiation is the incoming direct solar energy incident on a surface *at right angles to the solar beam.* (All other definitions refer to the radiation falling on a *horizontal* surface.) It does not include any diffuse radiation.

  Diffuse solar radiation is that part of the incoming solar energy incident on a horizontal surface shielded from direct radiation from the sun.

  Reflected solar radiation is that portion of the total solar energy reflected from the ground (and atmosphere) upwards without a change of wavelength. The ratio of this to the total incoming solar radiation is the *albedo* of the surface, more correctly known as the *reflection coefficient* of the ground (which must be differentiated from the surface's *reflectivity*, the fraction of the total incoming solar radiation reflected at a specific wavelength). For a vegetative cover of short crops, for example, the coefficient is about 0.25.

  Net radiation is the difference between all incoming and all outgoing radiation fluxes at all wavelengths (solar and terrestrial) on a horizontal surface. More details are given below under 'Net radiation'.

Total incoming radiation is the sum of all incoming solar and terrestrial radiation.

Total outgoing radiation is the sum of all outgoing solar and terrestrial radiation.

Incoming long-wave radiation is that radiated downwards from the atmosphere and clouds.

Outgoing long-wave radiation is that radiated upwards from the ground and atmosphere and clouds.

Photosynthetically active radiation (PAR), covering the band from 0.4 to 0.7 μm, is the spectrum used by plants for photosynthesis.

Ultraviolet (UV) radiation extends from about 0.1 to 0.4 μm and is subdivided into three narrower bands designated A, B and C.

Daylight illumination refers to the visual quality of the radiation in terms of the spectrum to which the human eye responds, which is from 0.38 to 0.78 μm, peaking at 0.55 μm.

Net radiation

Net radiation was defined above as the difference between all incoming and all outgoing radiation at all wavelengths (solar and terrestrial) on a horizontal surface. This is the total energy available for powering evaporation (*the latent heat flux*), for heating the atmosphere (the *sensible heat flux* – 'sensible' in that we can sense the air temperature), for warming the ground (*soil heat flux)* and for photochemical processes (photosynthesis). The soil heat flux represents a loss of energy during the day and a gain at night, averaging near zero over 24 hours, although there can also be a small seasonal trend that can affect evaporation rates slightly. Soil heat flux is very small for forests but can be high for short vegetation or bare ground. The relative amounts of net energy apportioned to evaporate water or heat the air depend on the nature of the surface – being different for bare soil, tall or short vegetation, open water or snow. It also depends on the amount of water available for evaporation, either free water or water within the soil or available to plant roots. Under some conditions, when water is particularly plentiful, some energy may also be drawn from the warmed air to power the evaporation process, in which case the sensible heat flux is negative.

Structure of the atmosphere and its terminology

Before considering how clouds form it is necessary to describe the structure of the atmosphere, something totally beyond the Greeks' comprehension. Even in the seventeenth century, little about it was appreciated.

Figure 3.4 shows the various layers into which we have divided and named the atmosphere. The caption explains all the relevant details.

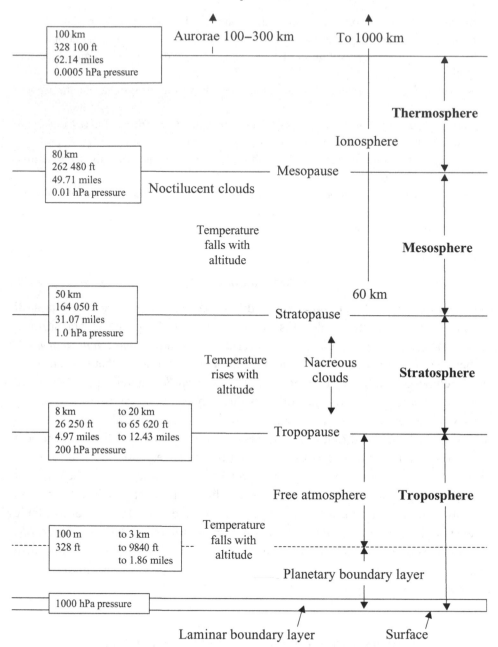

Figure 3.4 This figure shows the zones into which the atmosphere has been divided and the names used to describe the several layers. The drawing is not to scale, particularly the laminar boundary layer, which is only a few millimetres thick, and the planetary boundary layer, which might only be a few hundred metres deep. Most clouds are contained within the troposphere, which extends up to 8 km at the poles and 20 km in the tropics, but a few cirrus and rarer types can occur in the stratosphere and occasionally even higher in the mesosphere (see Chapter 4).

Dispersion of water vapour into the atmosphere

It was noted earlier that provided water vapour (WV) can be dispersed away from the water source from which it is originating, evaporation is not impeded. In the free atmosphere, the dispersion of evaporated water molecules away from the liquid water surface is a three-stage process. By the same processes, warmed air is also diffused away into the atmosphere.

Dispersion by molecular diffusion

In the first stage of dispersion of water vapour away from its source, we need to look initially at processes occurring very near to the surface of the water or wet object. Close to all objects, including the ground and plants, there is a thin layer of air, just a few millimetres thick, investigated early in the twentieth century by Ludwig Prandtl and Theodore Karman, known as the *laminar boundary layer*, which adheres strongly to all surfaces. It had been found by the Irish researcher from Belfast, Osborne Reynolds, that transition from laminar (smooth) to turbulent flow past an object (fluid in pipes in the case of Reynolds) could be calculated from the velocity, viscosity and density of the fluid and the dimensions of the object that the fluid was flowing past. The Reynolds number, which determines the state of flow, is expressed as:

$$Re = V_s \cdot L \cdot \rho/\mu$$

where V_s = mean velocity of flow
L = characteristic length of the object
ρ = density of the fluid
μ = viscosity of the fluid

Turbulence cannot penetrate the laminar flow region, and within this thin layer heat and water vapour are transferred entirely by molecular diffusion, the rate of transfer being proportional to the gradient of the properties. This is such a quantum-scale process that if it was the only form of diffusion away from the surface then in the case of temperature, for example, the daily cycle would vanish at four metres above the ground, and at two metres height the maximum temperature would occur around midnight. Fortunately there are more powerful processes at play.

Dispersion by forced convection

Convection is the transfer of heat through the movement of a fluid, in our case the atmosphere (in which the movement is predominantly vertical). It takes two forms.

In *forced convection* the movement is produced by the wind, particularly its turbulence, while in *natural* or *free convection* the movement is brought about through the buoyancy effects of rising convective cells. *Mixed convection* is a combination of forced and free convection. The term *advection* refers to the transport of air and its properties by motion in *any* direction, such as that caused by the horizontal wind. Thus heat and WV can be advected sideways as well as being convected vertically.

All land and sea surfaces are to varying degrees aerodynamically 'rough', and wind blowing horizontally across them suffers drag and is slowed down. The effect extends upwards to varying heights depending on the nature of the surface, but generally it can be assumed that up to a height of about 20 metres, windspeed increases logarithmically, thereafter increasing more slowly until it reaches the upper winds unaffected by the ground. Surface roughness also gives rise to *frictional turbulence* – random packets of air of varying size appearing briefly in a chaotic and transitory way which cannot be predicted but which nevertheless has some consistency. It occurs at many scales from centimetres to tens or hundreds of metres. Over the last decade, the availability of ultrasonic three-dimensional anemometers has allowed turbulence to be studied in detail. One such study (Shuttleworth 1979) was concerned with developing an instrument to measure evaporation directly by the *eddy correlation* method using an ultrasonic anemometer combined with a rapid-response IR humidity sensor.

This turbulent movement of the air takes over from molecular diffusion and carries moisture and heat away vertically from the surface on a large scale. It is the most powerful of the three processes by which water vapour and warmed air are propagated away from the surface.

Dispersion by free convection

Even in the absence of wind, provided there is a fall of temperature with increasing height, transfer of WV will take place through free convection, packets of warmed moist air rising up through the colder surrounding atmosphere due to their buoyancy. How far the warm air rises depends on the atmospheric conditions, but free convection acting alone would ensure that water vapour and warmed air were carried away from their sources even in the absence of wind. Free convection is also a vital element in cloud formation and so is dealt with in more detail in the next chapter.

Usually both forced and free convection occur together as mixed convection, wind effects normally being uppermost, free convection acting as a modifying influence. There is no need here to give a mathematical analysis of the processes involved but for readers who would like to know more, a full mathematical treatment is given by Shuttleworth (1979).

Lapse rate

The atmosphere is warmed mostly from the ground up, firstly by molecular diffusion through the laminar boundary layer and then by forced and free convection, but due to differences of slope, soil type and vegetation, solar radiation warms the ground unevenly, causing slight local differences of air temperature. When warmed, air becomes less dense and starts to rise through the cooler and denser air above it. At places where there are hot spots on the ground, bubbles or packets of more buoyant air, or *thermals*, form and rise more quickly than the surrounding slightly less warm air. Although after they have risen a few metres above the ground their heat supply is usually cut off, provided there is little mixing with the surrounding air, they conserve their heat and continue to ascend. As they rise, they cool; why?

The first law of thermodynamics

This law can be simply stated as 'energy is conserved'. In our case we are interested in energy in the form of heat (the temperature of the air) and of the mechanical work done by the air on its surroundings as the gas expands. In practical terms this means that as a warm packet of air rises and expands it does work on its surroundings, and in so doing some of its heat energy is converted into the physical work of expansion. By the first law of thermodynamics these must balance, and so the heat energy is reduced, cooling the air packet, the air molecules moving slightly more slowly. For the same reason subsiding air, increasing in pressure, warms, work being done on it by the surroundings. The amount by which the temperature changes can be quantified by the gas equations.

The gas laws

Boyle's law says that for a fixed mass of gas, kept at constant temperature, the volume is inversely proportional to the pressure. This can be expressed as:

$$P \propto 1/V, \text{ or } P \cdot V = constant, \text{ or } P_1 \cdot V_1 = P_2 \cdot V_2$$

Charles's law says that for a fixed mass of gas, kept at constant pressure, the volume is directly proportional to the temperature measured on the absolute scale (degrees Kelvin), this being expressed as:

$$V \propto T, \text{ or } V/T = constant, \text{ or } V_1/T_1 = V_2/T_2$$

The *pressure law* says that for a fixed mass of gas, kept at constant volume, the pressure is directly proportional to the absolute temperature, expressed as:

$$P \propto T, \text{ or } P/T = \text{constant}, \text{ or } P_1/T_1 = P_2/T_2$$

The *ideal gas equation* combines these three laws for a fixed mass of gas, giving:

$$P_1 \cdot V_1/T_1 = P_2 \cdot V_2/T_2, \text{ or } P \cdot V = \text{constant} \cdot T$$

The constant is controlled by the amount of gas present, which can be measured in three ways: its mass, the number of moles of gas (n), or the number of molecules of gas present (N). In the latter case the ideal gas equation can be written $P \cdot V = N \cdot k \cdot T$, where k is the Boltzmann constant.

Adiabatic lapse rate

As air does not conduct heat very well, and provided there is little turbulent mixing of the air packet with its surroundings, the cooling due to ascent is largely constrained within the packet or bubble. (Where heat does not enter or leave a system, the process is termed *adiabatic*. When heat is exchanged it is called *diabatic*, although the expression *non-adiabatic* is normally used.) Provided the dew point is not reached, the air cools at between 6 and 10 °C km^{-1} (the *dry adiabatic lapse rate*). If the thermal cools to the same temperature as its surroundings before reaching dew point, it stops rising, and that is the end of the matter; nothing is to be seen. If, however, the dew point is reached during ascent, cloud droplets form. But when condensation occurs this releases the latent heat that was acquired when the water was evaporated on the ground before its ascent. This in turn warms the thermal and so reduces its rate of cooling. Typically the *wet adiabatic lapse rate* is between 4.5 and 5.0 °C km^{-1}. This warming can even accelerate the rate of climb of the thermal, or at least allow it to continue rising further than it would have done if the dew point had not been reached, thereby extending its ceiling. How far the thermal rises, with or without condensation, depends on the conditions of the atmosphere through which it is rising (see 'Stability and instability', below). It is entirely through this cooling by expansion during ascent that most clouds are formed. Just a few are cooled by other processes (see next chapter).

Inversions

In the troposphere (Fig. 3.4), temperature usually falls with increasing height at a lapse rate depending on the conditions, as explained in the preceding section. However, the lapse rate is not necessarily constant all the way up since the atmosphere is often not homogeneous throughout its depth and can contain several layers with different temperature and water vapour profiles. In some layers, the temperature can actually increase with height, and this is known as an *inversion*. Such conditions put a limit on the height to which thermals can rise, creating a virtual 'ceiling' through which they cannot climb further. When this happens the thermal spreads out sideways.

There is a permanent inversion starting at the tropopause, extending up through the entire stratosphere to about 50 km, this increase in temperature with height being due to ozone in that region being heated through absorption of UV radiation. The height of the tropopause (top of the troposphere) varies (Fig. 3.4) from 20 km at the equator to 8 km at the poles. The inversion that starts at these heights often prevents the further continuation of growth of the larger cumulus clouds, causing them to spread sideways and thereby generating the classic anvil shape of thunder clouds (see next chapter). The tropopause is not, however, an absolutely impenetrable ceiling, there being breaks in its continuity where its level changes abruptly, allowing mixing to occur. These discontinuities occur where there are large temperature gradients, in particular near to polar fronts and the subtropical high-pressure zone.

Stability and instability

If the air into which a thermal rises is warm or has a low lapse rate, a thermal rising through it will soon come to a stop. These conditions are called *stable*. (Inversions are layers of great stability.) If, however, the air is cold or there is a high lapse rate, a rising thermal can reach greater heights and if it also reaches dew point, resulting in condensation, the heat thus gained will continue the thermal's rise further. In this way thermals may rise until they meet the inversion at the tropopause. These conditions are called *unstable*. Cloud-base height is the height at which the dew point is reached, while the overall height of a cloud is dependent on the stability of the atmosphere.

With this background we can now look at how clouds form.

References

Meteorological Branch, Department of Transport, Canada (1962). *Manual of Radiation Instruments and Observations*. Circular 3812 INS 117, Manual 84.
Monteith, J. L. (1975). *Principles of Environmental Physics*. London: Edward Arnold.
Shuttleworth, W. J. (1979). *Evaporation*. Institute of Hydrology Report 56. Wallingford: Institute of Hydrology.

4

Cloud formation

Around 1940 *cloud physics* appeared as a new meteorological discipline, concentrating mostly on *cloud microphysics*, which is the study of how the cloud particles form. Mason (1957) soon realised, however, that cloud microphysics was only part of the subject and that another, which can be termed *cloud dynamics*, needed to be added. This concerned the motion of the atmosphere involved in cloud formation, from the small scale of a kilometre or less to over 1000 km. Cloud dynamics was much more difficult to investigate than microphysics because of its large scale and because of the lack of suitable tools at the time. But with the advent of radar, specialised aircraft, satellite imagery and computers with numerical modelling techniques it has become possible to identify most of the important air movements associated with all cloud types. However, it is not possible in one chapter to include the very detailed knowledge now available on this, and just sufficient will be said to provide a background to the formation of clouds. The formation of cloud droplets or ice crystals and the formation, from them, of raindrops, hail and snow are dealt with in the next chapter.

Cloud classification and identification

In the uncomplicated situation, the identification of clouds is straightforward, but it can be quite difficult if there are several types co-existing at the same time. Also one type can change into another, leading to halfway stages possibly causing one to be confused with another. And some types of cloud can develop by various different paths, making identification a skill that requires time to develop. Our main concern, however, is how clouds form, not with the minutiae of how to identify them, although, to discuss formation, some basic knowledge of identification is necessary.

Readers interested in delving deeper can look at the *International Cloud Atlas* compiled by WMO, volume I (text) and volume II (photographs) (WMO 1975, 1987). This is the primary reference work for professionals, and looking through its

Table 4.1 *The ten principal cloud types.*

Genus (type)	Abbreviation
Cumulus	Cu
Stratocumulus	Sc
Altocumulus	Ac
Cirrocumulus	Cc
Cumulonimbus	Cb
Stratus	St
Nimbostratus	Ns
Altostratus	As
Cirrostratus	Cs
Cirrus	Ci

127 photographs (together with more from aircraft) illustrates the complexity of the combinations, mixtures and hybrids of cloud forms. The UK Met Office *Observer's Handbook* (1982) gives a briefer set of instructions for cloud identification. One of the foremost cloud experts, R. S. Scorer (1972) produced a complete cloud colour encyclopaedia, while popular books, such as that by Dunlop (2003), are aimed at amateurs and can be very useful. The Royal Meteorological Society magazine *Weather* has a selection of photographs each month, usually with one on the cover, while in 2003 there was a special issue on clouds and Luke Howard (*Weather* 2003). There are also websites with cloud information, such as those of the UK Met office (www.metoffice.com), RMS (www.rmets.org) and AMS (www.ametsoc.org). Books that are helpful in understanding the mechanisms leading to cloud formation are those by Scorer and Verkask (1989), Houze (1993), Scorer (1994) and Bader *et al.* (1995).

Cloud names

As with plants and animals, the classification of clouds is indicated by Latinate names, and they are similarly divided into genus, species and variety. Height of the base of the cloud is another important defining feature. The three basic forms of cloud are *cumulus* (heaped), *stratus* (layered) and *cirrus* (feathered), first named by Luke Howard in 1802 (Howard 1804, Hamblyn 2001). These names are still used although Howard also included a fourth type, *nimbus*, meaning rain-bearing. This is not now used on its own, but does appear as part of the name of the two main rain-bearing clouds, *nimbostratus* and *cumulonimbus*. Cloud types are given abbreviations, such as *Cu* for cumulus (Table 4.1). A further Latin term, *alto*, for height, is also used in conjunction with stratus and cumulus to indicate middle-height clouds.

Table 4.2 *Cloud species.*

Species	Abbreviation	Description
calvus	cal	Smooth top
capillatus	cap	Fibrous; striated
castellanus	cas	Turrets
congestus	con	Growing vertically; sprouting
fibratus	fib	Nearly straight; no hooks
floccus	flo	Tufts
fractus	fra	Ragged shreds
humilis	hum	Flattened
lenticularis	len	Lens-shaped
mediocris	med	Moderate depth
nebulosus	neb	Featureless thin layer
spissatus	spi	Denser cirrus
stratiformis	str	Horizontal layer
uncinus	unc	Hook-shaped

It is useful to add an eleventh type – fog – to those listed in Table 4.1, which by definition is any cloud that extends down to ground level. It is not included in the standard international list since fog is covered under 'visibility' in the international reporting codes. Also if an observation is made on a mountain, what looks like fog to the observer on the spot might be reported by another observer lower down as stratus or orographic cloud (see later). For these reasons it is more useful to use the definition given by Houze (1993) that true fog is produced as a *direct consequence* of the air being in contact with the ground.

In addition to the ten main cloud types, there are various subtypes, or *species*, as shown in Table 4.2, with three-letter abbreviations in lower case, and further subdivisions into cloud *varieties*, *supplementary features* and *accessory clouds*, as in Tables 4.3, 4.4 and 4.5. I hesitated before including these since they will be referred to hereafter only passingly, but it is useful for the sake of completeness.

Cloud height

The method of dividing the atmosphere vertically into three layers or *étages* (Table 4.6) was introduced by the French naturalist Jean-Baptiste Lamarck. The height refers to the height of the base of the cloud not to its overall height and so some clouds, such as cumulonimbus, are listed as low clouds although they can extend up to the tropopause. Different authorities give slightly different ranges, the ones given here being WMO's. Feet have been used instead of metres since aviation uses feet internationally, and so they will be more familiar to readers (1 foot = 0.305 m). The

Table 4.3 *Cloud varieties.*

Variety	Abbreviation	Description
duplicatus	du	Several layers
intortus	in	Irregularly curved or tangled
lacunosus	la	Thin cloud with regular holes
opacus	op	Masks sun
perlucidus	pe	Broad patches with small spaces
radiatus	ra	Bands, apparently converging
translucidus	tr	Sun shows through
undulatus	un	Sheets with parallel undulations
vertebratus	ve	Like vertebrae

Table 4.4 *Supplementary features of clouds.*

Name	Abbreviation	Description
arcus	arc	Arched cloud
incus	inc	Anvil cloud
mamma	mam	Hanging pouches
praecipitatio	pre	Precipitation reaching ground
tuba	tub	Funnel cloud
virga (fallstreak)	vir	Droplets falling under cloud

Table 4.5 *Accessory clouds.*

Name	Abbreviation	Description
pannus	pan	Shreds of cloud
pileus	pil	Cap cloud
velum	vel	Veil

heights vary with latitude and those given are for mid latitudes. While there is no height difference for the lower clouds, for cirrus the polar range is 10 000–25 000 feet while for the tropics it is 20 000–60 000 feet. The one set of heights shown in Table 4.6 is sufficient for our purposes.

Cloud observation from satellites

The generation of images of clouds was one of the main purposes of the first weather satellites. These images are described fully in Chapter 11 and it is best to leave the matter of automatic cloud identification from satellites until then.

Table 4.6 *Height of cloud base.*

Genus	Étage	Cloud base (feet)
stratus	Low	Surface–2000
stratocumulus		1000–4500
cumulus		1000–5000
cumulonimbus		2000–5000
nimbostratus	Middle	Surface–10 000
altostratus		6500–20 000
altocumulus		6500–20 000
cirrus	High	20 000–40 000
cirrostratus		
cirrocumulus		

1 foot = 0.305 m.

Sufficient here to say that while it is possible to go part way towards measuring cloud cover and classifying genera automatically from satellite images, and while an experienced observer can deduce much about cloud cover and type simply by looking at satellite images (Scorer 1994), it is not yet possible to replace the ground-based observer looking up at the sky and making judgements as to the subtleties and detail of the cloud cover overhead. There is little doubt, however, that automatic identification will eventually be realised as algorithms are developed and image resolution is increased. But for the present the human observer on the ground remains the only sure way of identification.

Units and terms

Total cloud amount (or *total cloud cover*) quantifies the fraction of the celestial dome covered by all types and levels of cloud. Cover is estimated to the nearest eighth, expressed in *oktas,* 0 representing no clouds at all, 8 indicating complete cover. A record of 9 oktas is used to show that an observation was not possible due, for example, to fog or falling snow. *Partial cloud amount* is the amount of sky covered by each type or level of cloud. The sum of partials can, therefore, be greater than 8 oktas, since there is usually some overlap at different levels. An instrument recently appeared on the market which senses the 'nebulosity' of the sky using four infrared sensors directed at the sky north, south, east and west.

Cloud base or *cloud height* is the height of the base of the cloud above the ground at the observation site, expressed in metres, or for aviation applications in feet. It is defined (by the WMO) as 'the lowest zone in which the type of obscuration perceptibly changes from that corresponding to clear air or haze to that corresponding to water droplets or ice crystals'. It is important to differentiate between cloud *height*, that is the height above the local ground, and cloud *altitude*, the height above mean sea level. This is usually done by saying *height above ground level*. Clearly this difference is important in aviation.

Cloud formation by advection and radiation

Stratus and fog

Stratus is the lowest-level cloud, its base rarely being higher than 2000 feet (0.75 km) and often reaching down to ground level. It is also the odd one out in that it does not necessarily depend on cooling by ascent for its formation, the most common cause of stratus being when warm moist air is carried by a very light wind over a cold land or sea surface. This produces cloud at ground level, known as *advection fog*, which is synonymous with *stratus*. (As defined in the previous chapter, 'advection' is the transport of air sideways rather then vertically by convection.)

If the wind is stronger, there will be more turbulence, mixing the air more extensively, and the cloud will form some distance above the ground, although still low, perhaps obscuring the tops of buildings or hills. Such cloud or fog is usually composed of water droplets forming a grey or blue-grey blanket, but if it is very cold there can be some ice crystals as well. Even though grey from below or from within, from above (Fig. 4.1) it can be brilliant white, looking like snow. It can also, on occasions, be thin enough for the sun or moon to be seen though it, and sometimes this can produce a corona or, if the cloud contains ice, a halo. (A *corona* is produced by water droplets and consists of a central bluish-white 'aureole' around the sun or moon (Fig. 4.1, inset), possibly with an outer coloured ring. A *halo* is produced by clouds formed of ice crystals and takes the form of a ring or part-ring of light around the sun. Most commonly the ring has a radius of 22° centred on the sun. This arises due to sunlight passing into one and out of another face of hexagonal ice crystals, which have faces at 60° to each other (see Fig. 5.2d,e; the exact mechanism is demonstrated at www.atoptics.co.uk). Fainter rings can also occur at 46° radius. Mock suns, bright spots of light at the same height as the sun, located on the 22° ring, and other effects, are also quite common. Coronas and haloes are more common than is thought but are just not noticed because people rarely look at the sky in the direction of the sun, shielding it with a hand or other object. For this reason they are more noticeable around the moon.)

Jonas (1994) explains that the formation of fog results from the mixing of two subsaturated volumes of air at different temperatures – the air advecting in and the air already present over the ground. While each volume of air alone is

Figure 4.1 Stratus is the lowest-level cloud, its base often reaching down to ground level, as here, filling a valley near Geneva. In this form, cloud is also known as valley fog. Inset: (left) the full moon through thin stratus with a corona; (centre) stratus forming in a shallow valley in November in the UK; (right) the sun through thin stratus.

below saturation at its particular temperature, when mixed, the combined volume at the combined temperature becomes saturated, causing condensation and thus cloud formation. This is because of the shape of the saturation mixing ratio curve (Fig. 3.2). In a reverse form of this process, if cold air advects in over warm water, a shallow layer of *steam fog* can form, usually a metre or so in depth. Again the process is one of the mixing of two volumes of air at different temperatures and degrees of saturation, the combined product becoming saturated.

Radiation fog forms when the sky is clear, allowing long-wave radiation to be lost from the ground to space, thereby cooling it. There needs to be a light wind to ensure that the cooling of the ground is transferred to the layers of air close to it. This type of fog can form very quickly and can reach a thickness of 50 to 500 feet (15–150 m), possibly 1000 (305 m). It never forms over the sea or large lakes since, unlike the ground, they do not cool quickly after sunset. The fog clears when heated by the sun or if the wind increases, mixing the saturated air with drier, warmer air above. A different, very shallow, radiation fog forms when the lower few metres of air itself (rather than the ground) cools by radiation. If the drops of fog are supercooled they form rime ice when they come in contact with objects (more on rime in the next chapter).

Smog or *smoke fog* arises when the air contains smoke particles which act as nuclei for cloud droplets to form on (see also next chapter), producing droplets smaller than pure water fog. I remember such fogs well from my childhood in the north of England. It was very persistent and unpleasant. Fortunately it has now gone, thanks to the Clean Air Acts. Before these Acts, people heated their houses by burning coal, steam trains burnt coal and factories burnt coal. Smoke was everywhere and the buildings were black while sheep in the uplands nearby were grey.

Yet another type of fog, *ice fog*, or *diamond dust*, occurs when the temperature is low enough to cause the fog droplets to freeze into very small ice crystals. These glitter in the sun, do not reduce visibility and produce haloes: an altogether more attractive phenomenon than smog.

Stratus and fog are the odd ones out, the great majority of clouds forming only through uplift of some form. Stratus, however, can also form by gentle uplift. This occurs principally at the leading edge of a warm front before it thickens into nimbostratus, or when moist air is forced to rise when the wind meets a mountain or hill, the tops becoming covered by stratus (or other) cloud if the dew point is reached (Orographic clouds: see below). Stratus can also develop when stratocumulus (see later) lowers or its subdivisions are filled in.

Because stratus is usually very thin, little precipitation is produced – although there can be light drizzle or a dusting of snow or ice crystals.

Clouds formed by convection

In the previous chapter it was shown how, when the ground is warmed by the sun and water is evaporated, moist thermals ascend through the atmosphere. As they rise they cool due to expansion, and if the dew point is reached the vapour condenses and clouds form. Let us now look at the clouds formed by this process.

The most familiar clouds to result from buoyant ascent are the cumulus family, which occur in a large variety of sizes and forms. A good illustration of this process in action can be seen by observing the warm, moist air from a power station's cooling towers turn into small cumulus clouds downwind of the station. Here the source of heat and water vapour is more apparent than in the case of natural thermals, and the mechanism is seen more clearly (Fig. 4.2). As a family, cumulus are detached, very dense, with sharp outlines growing to form mounds and domes, the larger clouds usually described as resembling cauliflowers. The sides exposed to the sun

Figure 4.2 Rising moist warm air from a power station's cooling towers illustrates the process of cumulus cloud formation. Here the source of thermals is obvious but it is the same process as is involved in natural thermals caused by the sun warming the ground and evaporating water. Here the only clouds are those from the power station, the atmosphere being fairly dry and stable.

are brilliant white, but the other sides can be dark, as are their bases, which are normally flat, at the condensation level.

Cumulus humilis and mediocris

If there is an inversion (see previous chapter) slightly higher than the condensation level, flattened cumulus clouds known as cumulus humilis form (Fig. 4.3). They tend to occur under anticyclonic conditions when slowly descending (and so warming) air restricts upward growth. These clouds typically form at a height of 3300 feet (1 km) or less and have a width of up to 1 km. They produce no precipitation. The smaller cumulus clouds are short-lived, rarely lasting more than 15 minutes. It is only necessary to watch a small patch of cloud, slightly separated from the edge of a cumulus cloud, to see it evaporate within minutes (Fig. 4.3, inset). When there is moderate vertical growth the cloud is known as cumulus mediocris (Fig. 4.4).

Cumulus congestus

When heating is very strong and the atmosphere is unstable (see previous chapter), the clouds grow bigger and become 'heaped' and more cauliflower-like. They are then known as cumulus congestus. Although the base is in the lowest étage, they can reach heights and widths of 2–3 km and often lean downwind because the windspeed

Figure 4.3 Natural thermals produce cumulus clouds, as here on a summer's day in the UK. Due to a fairly low inversion just above the condensation level, the clouds are limited to a small vertical extent, producing cumulus humilis. The short life cycle of these clouds is illustrated by the three lower photographs of a small patch of cloud taken at one-minute intervals. It is easy to appreciate the puzzlement of the Greeks when they saw a cloud disappear (or appear) 'out of thin air'; their conclusion was that air can turn into cloud and back again.

is usually greater at higher altitudes due to wind shear (Fig. 4.5). Freezing does not occur and so the tops stay hard-edged. When ice forms, clouds take on a fibrous appearance, as in the next type. In cool climates cumulus congestus only occasionally produce precipitation, but in the tropics they produce plentiful amounts.

Cumulonimbus

If cumulus congestus clouds continue to grow they can evolve into the largest of the cumulus family, indeed into the largest of all clouds in vertical extent,

Figure 4.4 If an inversion is higher, and the air less stable, clouds can grow to a greater depth, resulting in cumulus mediocris.

Figure 4.5 Given more unstable air, cumulus clouds can grow yet taller, as here over the New Forest in the UK in October, the resultant cumulus congestus leaning to the left due to wind shear, windspeed increasing with height.

Figure 4.6 If there is no restraint on upward growth, cumulus congestus can grow into the larger cumulonimbus until it reaches the tropopause. As the height increases a point may be reached when freezing starts to occur, whereupon the sharp edge of the cauliflower-like cloud begins to soften and become fibrous as in the lower right photograph. If an inversion is met, the cloud spreads out sideways, forming the familiar anvil top as in the upper photograph, taken from a height of around 35 000 feet over a tropical ocean. A cap cloud (pileus) has formed where the rapidly rising updraught has forced itself through the inversion for a short distance. The lower left figure is of two winter cumulonimbus clouds in the UK where the tropopause was much lower.

cumulonimbus. These can take several forms, the larger ones producing hail and thunder and, at times, tornadoes.

Single-cell storms

These are the smallest and simplest of the cumulonimbus clouds (Fig. 4.6) their growth from cumulus humilis to congestus taking about 20–30 minutes followed by a further similar period to build to cumulonimbus, thereafter subsiding over the next 30 minutes to two hours. Only one main updraught is involved. Nevertheless, if large

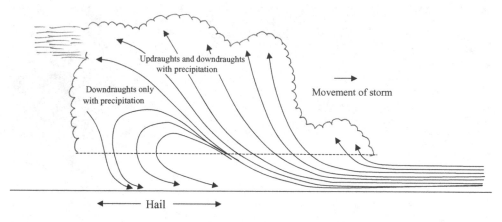

Figure 4.7 Cumulonimbus clouds often occur in multicell form, looking like one
larger single cloud while in reality they are several cells one following the next as
they grow to be the main cell, then decaying. Cells that are in the early stage of
formation have vigorous updraughts with rapidly forming rain and graupel while
mature cells have both upward and downward currents, the downdraughts being
accompanied by heavy precipitation. As the cells finally decay, after their relatively short lifetime, only downdraughts occur, with continuing precipitation. The
drawing is based on Houze (1993).

enough, single-cell storms can produce hail and thunder (the processes leading to
hail formation, and the charging processes that result in lightning, are considered in
Chapters 5 and 6 respectively). The updraught, which can range from 1 to $10 \, \mathrm{m \, s^{-1}}$,
tends to draw warmed air preferentially from in advance of the cloud's direction of
travel, the cloud as a whole being carried along by the environmental wind, giving
the impression to someone standing on the ground that the cloud 'travels into the
wind'.

If the top of the cloud reaches an inversion, typically the tropopause, it flattens
out, unable to penetrate into the stable stratosphere above, although if growth is
particularly vigorous the updraught can push through slightly into the stratosphere
(Fig. 4.6) producing a *cap cloud* or 'pileus'. If freezing occurs near the top of the
cloud, the rounded clear-cut profile of congestus become fibrous and soft-edged –
always a sign of freezing. If there is wind shear, due to an increase of windspeed
with altitude, the cloud may be drawn out to produce the familiar anvil-shaped
top. Single-, multi- and super-cell clouds can all reach up as far as the troposphere
while having bases near to the ground. Their horizontal extent can be in tens of
kilometres.

Multicell storms

Very often several convective cells will cluster together as one multicell storm
(Fig. 4.7), each individually with the short lifetime of a single cell, but as a group

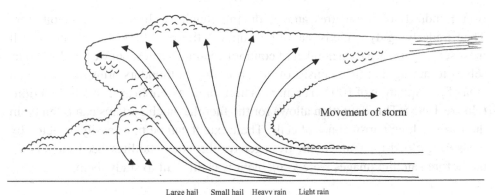

Large hail   Small hail   Heavy rain   Light rain

Figure 4.8 The largest cumulonimbus cloud is the supercell, a very large and violent single-cell thunderstorm made up of one updraught and one downdraught. The mechanics of these cells is, however, complex, involving rotating updraughts, resulting not only in lightning but also in very large hail and sometimes tornadoes. The drawing is based on Houze (1993).

able to last several hours as new cells grow and replace the decaying cells, giving the impression that it is one larger, longer-lived cloud while in reality it is a constantly renewed system of many single cells, each taking its turn to be the dominant main cell. Their overall size can be up to 10 km.

Downdraught are prominent characteristics of thunderstorms and can become very intense, when they are know as *downbursts*, defined as an area of strong wind produced by a downdraught over an area up to 10 km in horizontal extent. *Macrobursts* are downbursts over an area greater than 4 km lasting 5–30 minutes while *Microbursts* cover an area less than 4 km and last 2–5 minutes. The main concern with the latter is in aviation, and air accidents have resulted from them; to warn of such conditions many airports now have instrumentation in place. As with most cloud dynamics, air movement involved in downbursts is complex and does not warrant too close a look here.

Supercell storms

The much rarer supercell thunderstorm consists of just one very powerful giant updraught of 10 to 40 m s^{-1} and one downdraught (Fig. 4.8). Radar has thrown new light on the internal structure of these, and multicell, storms but the processes are basically similar to those of the single-cell form. The detail of the air movements within these clouds, often involving the generation of tornadoes, is too complex to go into here.

Mesoscale convection

In temperate regions we are familiar with isolated thunderstorms but in the tropics they occur in large clusters, termed *mesoscale convective systems*, extending over hundreds of kilometres and producing much of the earth's precipitation. Their underlying dynamic is that of buoyant air, thoroughly dealt with above. All mesoscale systems have a number of common characteristics, but we will limit ourselves to noting that they consist of 'an ensemble of thunderstorms producing an area of precipitation of 100 km or more in horizontal extent in at least one direction' (Houze 1993). This definition allows for the fact that such systems can often be in the form of long narrow lines of cells. Their extent can be most easily detected by measuring cloud-top temperature from geosynchronous satellites (see Chapter 11), but before satellite images were available it was difficult to study them.

Hurricanes, typhoons and cyclones

Tropical cyclones are known as *hurricanes* in the North Atlantic and eastern Pacific, *typhoons* in the western Pacific and *cyclones* in the Indian Ocean and Coral Sea.

Due to the meeting of the easterly trades from the north and south along the equatorial belt known as the Intertropical Convergence Zone, and because of the high sea surface temperatures (SSTs) in these regions, a belt of thunderclouds exists around the globe. These sometimes group into clusters, and if the wind shear is low, allowing strong vertical growth, and the SST is more than 26.5 °C, the clouds can become very large, the latent heat released upon condensation powering their growth. If a low pressure forms at the centre of the cluster, with some circulation, it becomes known as a *tropical depression*, winds near the centre being from 37 to 62 km per hour. If the depression intensifies to the point where the windspeed increases to 62–117 km per hour it is known as a *tropical storm* and is given a name, the structure becoming more circular, resembling a hurricane. These storms can cause damage but mostly through heavy rain (120–240 mm). If the surface pressure drops further and the circulating windspeed becomes greater than 118 km per hour (75 mph), the storm becomes a *hurricane*. Hurricanes are rated according to speed on the Saffir–Simpson scale of 1–5. Around the eye of the hurricane, which is 20–50 km in diameter, is the eye wall, where most of the damaging winds occur, along with the highest rainfall. Bands of cloud and precipitation spiral outwards from the eye. Why hurricanes have eyes, and their general structure, are addressed by Pearce (2005) and Buontempo *et al.* (2006).

Hurricanes form in latitudes from 5 to 20 degrees north and south, but not on the equator (where there is no Coriolis force – a certain amount of vorticity is required to trigger them). Once formed they tend to move west, but always turn east eventually, although the factors affecting their movement are complex. This makes accurate prediction difficult, and empirical methods are often used. As far as we are concerned, however, hurricanes are another arrangement of large cumulus clouds and we will not digress to take a closer look at them, even though they can be

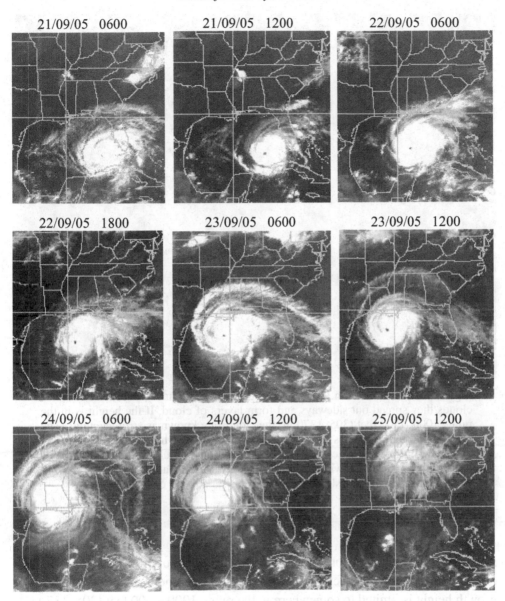

Figure 4.9 The path of hurricane Rita, September 2005, followed by the GOES E satellite.

dramatic and newsworthy, as witnessed by the events of 2005, when the category 5 hurricanes Katrina, Rita (Fig. 4.9) and Wilma caused massive destruction in the Caribbean and the USA.

It is necessary to go back slightly now to look at the clouds that form by convection if inversions prevent a rise beyond a certain height.

Figure 4.10 If there is an inversion preventing the continued ascent of cumulus clouds they spread out sideways and form layers of cloud. If the height is in the range 1000–4500 feet (305–1370 m) the clouds are known as *stratocumulus*. The individual rounded masses appear greater than 5° across when viewed at more than 30° above the horizon. The variety illustrated is stratocumulus undulatus. The rolls are at right angles to the wind and are caused by wind shear, the upper wind being faster than the lower, causing the rolls to rotate and move downwind.

Stratocumulus

If there is an inversion that prevents the continued ascent of ordinary small cumulus clouds to greater heights, they will spread out sideways and form a layer of cloud. If growth height is limited to somewhere in the range 1000–4500 feet (305–1370 m) the clouds so produced are known as *stratocumulus* (Fig. 4.10). Early in the day there can be large clear areas of sky between the individual cumulus clouds, but these gradually fill as cloud amounts increase.

It is also possible for stratocumulus to evolve when a sheet of stratus is broken up by convection. It is the most common form of cloud globally, and satellite images show that it occurs over widespread areas of the oceans. Although occasionally there can be trails of falling water droplets or ice crystals (fallstreaks or *virga*) they

Figure 4.11 Altocumulus can be formed by the same processes as stratocumulus – that is by an inversion causing a sideways spread of cumulus cloud, but in this type the inversion is higher (6500–20 000 feet, 2000–6000 m), placing the cloud in the mid level. The individual rounded masses appear less than 5° across when viewed at more than 30° above the horizon. The cloud illustrated is altocumulus undulatus, the rolls being produced by wind shear as in the previous figure.

usually evaporate before reaching the ground, and stratocumulus rarely produces much precipitation.

Altocumulus

Altocumulus has much in common with stratocumulus and can be formed by the same processes – that is by an inversion causing a sideways spread of cumulus cloud, but in this case the inversion is higher (6500–20 000 feet, 2000–6000 m) placing the cloud in the mid level (Fig. 4.11). Altocumulus can also form from altostratus, stratocumulus and nimbostratus as well as behind active fronts (see later), so their origins are diverse and not necessarily due to the sideways spread of cumulus, causing its appearance to be very varied with four species and seven varieties.

We will move on now from clouds caused by rising air, brought about by convection, to cooling caused by another quite different mechanism – rising due to the meeting of warm and cold air masses.

Extratropical cyclones

Just as the buoyant ascent of thermals produces cooling which can result in cumulus cloud formation, cooling also occurs when a warm mass of air meets a cold mass and rises over it, or when a cold mass of air moves in beneath a warm mass, lifting it. This process generates clouds of several types and occurs at *frontal depressions*, more correctly known as *extratropical cyclones*, but usually referred to simply as *depressions* (as they will be referred to hereafter). They are a common feature of mid-latitude climates and a source of most of the precipitation in these regions.

Although personalities do not figure greatly beyond Part 1 of the book, it is interesting to note that after a few years of observation, Robert FitzRoy (1863) proposed a model for depressions. Having spent much of his life at sea, earlier as Captain of the *Beagle* taking Charles Darwin on his voyage of discovery around the globe from 27 December 1831 to 2 October 1836, and later having promoted the use of mercury barometers aboard ship, he was ideally placed to develop these ideas. He also became the first Head of the UK Met Office, founded in 1854. A close observer of detail, as he had been on his journey with Darwin (although coming to totally different conclusions about what he saw), FitzRoy noted that depressions were usually the product of two air masses of different temperature and humidity, the warmer and humid coming north from subtropical latitudes, the colder and drier air descending from the polar regions. As a result of his observations and conclusions, FitzRoy was able to predict future weather and actually invented the term 'weather forecasting'. As an important contributor to science, FitzRoy is underestimated, being rather under the shadow of Darwin.

Sixty years later, Bjerknes elaborated on FitzRoy's original ideas and, having analysed many depressions, devised a new model that described their structure and dynamic processes in more detail. The model explained the types of cloud and precipitation that result from the cooling of the warm air as it ascends at both the warm and cold fronts. A few years later Bjerknes and Solberg (1922) observed that depressions usually go through a series of typical stages, starting as a small wave, triggered by wind shear, on the more or less straight interface between the warm and cold air masses. FitzRoy, and later Bjerknes and Solberg, also noted that an isolated single depression rarely occurs on its own, there normally being two, three or more, one in the wake of the other, all tending to move northeast (in the northern hemisphere). While movement northeastward is quite rapid when a depression first forms, it gradually slows and eventually stops as the front becomes increasingly, and finally completely, occluded (see below), giving rise, for example, to the persistent Icelandic and Aleutian lows.

Figure 4.12 shows a plan view of the sequence of events that occur as a depression develops, with a vertical cross-section through the troposphere (Fig. 4.13) showing how the warm air is forced aloft above the cold, giving rise to a set sequence of clouds. The figures sum up the work of FitzRoy, Bjerknes and Solberg and their drawings and are my reworking of their various diagrams. A look at actual surface pressure maps, however, often shows something much more complex than the simple situation illustrated in Figs. 4.12 and 4.13. Not only can there be several lows and highs with fronts associated with them, all interacting, but the fronts can be drawn out or exist as single isolated fronts.

This model has served well as the basis for weather forecasting since it was first proposed, but the development of weather radar in the 1950s (Collier 1996) and the availability of satellite images since the 1960s have shown that the cloud structure within depressions is more complex than it had been possible to establish visually from the ground (Collier 2003) – lacking the ability to see through intervening stratus cloud to what was happening above, or to see events beyond the horizon. Using the concept of ascent through 'conveyor belts', these new remotely sensed observations have made it possible to develop more complex models showing in greater detail the three-dimensional structure of the airflow within depressions, resulting in a better understanding of the generation of cloudbands and rainbands. However, all depressions are not the same, the exact configuration of the 'conveyor belts' having a strong influence on the clouds that form (McGinnigle *et al.* 1988). Nonetheless, for our purposes it is sufficient to use the classical view of FitzRoy, Bjerknes and Solberg and not to overcomplicate the matter by being too anxious to include every possible detail, which might confuse rather than enlighten.

With the free availability of satellite images on the internet, it has become simple to follow the progress of depressions. I use the UK Met Office site (www.metoffice.com), which gives hourly updated visible and IR images. The Met Office site also gives surface pressure charts with forward predictions, in six-hourly intervals, on which the fronts are marked. The satellite image of Fig. 4.14 illustrates the appearance of a typical depression.

Clouds formed at warm fronts

Once a depression has developed to the stage shown in Fig. 4.12, frame 4, the full range of clouds all exist at the same time, spread out over a few hundred kilometres along the line A–A, all moving gradually northeast (in the northern hemisphere), the front slowly 'occluding' as it travels and evolves, the warm air finally being lifted entirely from the ground with cold air beneath (an occluded front).

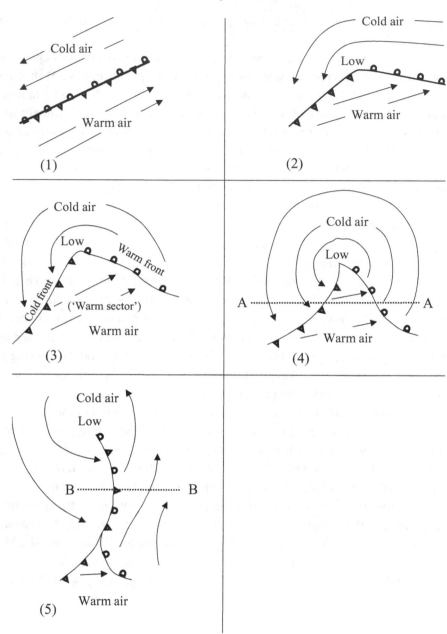

Figure 4.12 The formation of an extratropical cyclone or depression goes through well-established phases (although all are different in detail), starting with a mass of dry cold arctic air adjoining a moist maritime tropical air mass (1). A wave develops in the initially straight interface and once it has developed a low pressure forms, around which the air circulates, the warm air being forced to rise over the colder (2). This pattern then evolves progressively as shown in frames (3), (4) and (5). As the warm air is forced aloft and cools, clouds develop, the type depending on altitude. Figure 4.13 shows cross-sections through the fronts.

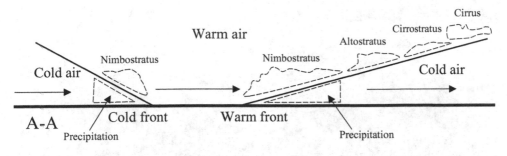

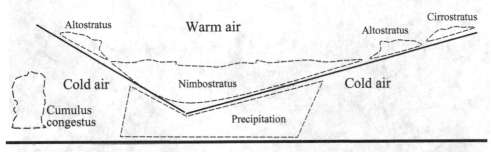

Figure 4.13 If a cross-section is taken along the line A–A (Fig. 4.12, frame 4) some of the warm air is still in contact with the ground (the 'warm sector'). Where the warm air has risen over the cold air, clouds form, the type depending on altitude. Where the cold air, at the cold front, forces the warm air aloft, clouds form of a similar type to those at the warm front, but in the reverse order. The drawing is not to scale, the slope of the warm front being about 1 in 100 to 150, the cirrus being around 800 km from the front, with 300 km of these experiencing precipitation. The slope at the cold front is steeper at about 1 in 50 to 75. Condensation into clouds releases latent heat and it is this that provides the energy to drive the circulating winds. Eventually (lower diagram) all of the warm air is lifted from the ground, as in the section B–B (Fig. 4.12, frame 5), and the front is occluded.

All of the clouds in this cycle are formed through the cooling of the warm air by uplift over the cold air beneath, the type of cloud depending on the altitude to which the warm air is raised.

Cirrus

Cirrus is the highest group of (normal) clouds, reaching up even into the stratosphere, and is the first type of cloud to be seen well in advance of an approaching warm front, as the cold air lifts the warm moist atmosphere increasingly higher.

Figure 4.14 This Meteosat image shows a depression southwest of the UK. By the time the depression had developed to this stage, the front was well occluded, as in Figure 4.12, frame 5. The cumulus clouds following behind the cold front in the unstable air are clearly visible. Note the signs of a fainter depression west of Ireland and south of Iceland. Copyright EUMETSAT/Met Office 2004. Published by the Met Office.

There are five species of cirrus (fibratus, uncinus, spissatus, castellanus and floccus) with four varieties (duplicatus, intortus, radiatus and vertebratus), the various species evolving one into another, starting with a rising 'head' of supercooled water droplets which then freeze. When the ice crystals meet a stable layer of atmosphere, such as the tropopause, ascent stops and the crystals begin to fall, being sufficiently large (and the SVP of ice low enough) for them to evaporate slowly enough to remain visible for long periods.

Figure 4.15 Cirrus is the highest of normal clouds, consisting of ice crystals, often falling from 'generating heads'. As the crystals fall they leave trailing filaments either straight, entangled or with hooked ends (uncinus), while the heads move along more quickly in the higher winds above. Inset is a short aircraft contrail, which is artificially produced cirrus. Short trails occur when the air aloft is dry, but as a warm front approaches the trails persist for longer and grow in length, a sign of an approaching depression.

If there are high winds at these altitudes, the 'heads' are carried along rapidly, the falling ice being left behind in the lesser winds below giving rise to a 'hook' at the top of the strand, cirrus uncinus, this form often being referred to as 'mares' tails'. When the cells producing the ice crystals have weakened and the ice particles falling away from the cells have become dissipated, they become longer and more exaggerated, giving a stringy or hair-like appearance (Fig. 4.15). Thus cirrus cloud is virtually all precipitation although the ice crystals rarely reach the ground. In the most advanced stage of development, cirrus fibratus, the 'hook' has gone, as has the original source of the ice crystals, the 'heads'.

Just prior to the appearance of cirrus clouds the contrails of aircraft are seen to be short due to the upper air being dry (Fig 4.15, inset), but as the warm front approaches nearer and cirrus develops, the contrails lengthen until they stretch from horizon to horizon.

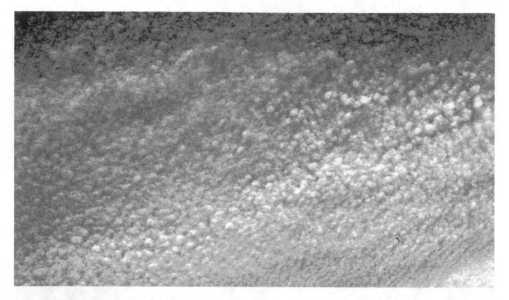

Figure 4.16   Cirrocumulus is the high counterpart of alto- and stratocumulus, being distinguished from them by the small size of the tufts and their thinness, the individual rounded masses appearing less than 1° across when viewed at more than 30° above the horizon. They often go unnoticed because of their small size and thinness. The contrast has been increased considerably in this photograph to make them clearly visible.

Cirrocumulus

The faint small pale white tufts of cirrocumulus (Fig. 4.16) are often not very visible due to their greater height, smaller-sized tufts and thinner depth compared with their counterparts, the lower, thicker and larger altocumulus and stratocumulus. While in temperate regions cirrus is mostly associated with warm fronts, in the tropics it is usually the remains of the anvil of cumulonimbus. As with all cirrus cloud they are composed mostly of ice crystals, any supercooled droplets being converted to ice quickly.

Cirrostratus

Cirrostratus is even more difficult to see than cirrocumulus, being a thin veil of ice crystals without any clear features, often only becoming noticeable as it thickens and starts to reduce solar intensity. It can form by the falling of ice from cirrocumulus tufts, the spreading out of cirrus streaks or directly through the rising of warm moist air as the warm front gets nearer (Fig. 4.17). As the warm front approaches, the cirrus gradually thickens, lowers and evolves into altostratus. The most noticeable feature of cirrostratus is it ability to produce halo phenomena and various arcs and mock suns.

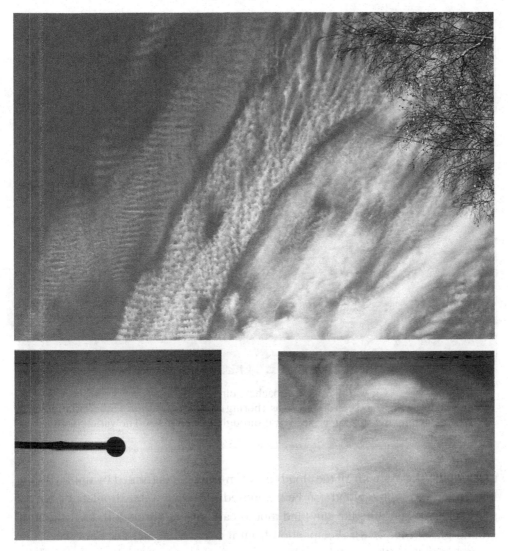

Figure 4.17 Cirrocumulus can evolve into cirrostratus and this can be seen happening in the right of the upper picture where the billows have spread into a layer of cirrostratus. Later the sky may become filled with a more extensive layer as in the bottom right image. In the bottom left image, a very thin layer of cirrostratus is beginning to form, made visible by looking in the direction of the sun, eclipsed by holding a disc in front to cast a shadow, otherwise it would go unnoticed. Under such conditions, a halo can sometimes be seen round the sun, as well as mock suns.

Altostratus

As the warm front advances further, cirrostratus cloud can lower and thicken into altostratus, with its base in the middle étage. It is an unexciting cloud showing little variation or texture and so is not divided into different species, although there are five varieties. It is light to dark grey, forming a featureless layer often covering

Figure 4.18   As a warm front approaches, cirrostratus may lower and become altostratus, which has been described as a boring cloud since it shows little variation. If thin enough, the sun can be seen as if through ground glass. The variety here is altostratus translucidus.

much of the sky. As with all the clouds at warm fronts, it is formed by uplift. Being lower than cirrus, the cloud can be composed of either water droplets or ice as snowflakes. When thin, the sun (and moon) can be seen through it, as if through ground glass (Fig. 4.18). Precipitation from it is minimal and does not reach the ground initially, although fallstreaks are common. But as it thickens, as the front approaches closer, it evolves into the much thicker nimbostratus.

Nimbostratus

The main rain-bearing cloud in mid latitudes, nimbostratus (Fig. 4.19) shows little structure and so, like altostratus, is not subdivided into species, nor does it have any special varieties. It is a dark, grey, heavy, layer cloud with a very ragged base and can become very deep, increasing in thickness as the warm air is progressively raised higher over the cooler air, its top sometimes reaching as far as the tropopause and so, like cumulonimbus, nimbostratus can also have its upper layers composed entirely of ice. It is never thin enough for the sun or moon to be seen through and

Figure 4.19 As the warm front finally arrives, the altostratus lowers into nimbostratus and precipitation starts which can continue for several hours.

always produces precipitation, which can continue for several hours because of its
wide extent. As the warm front finally passes, there may be a short period of low
stratus cloud, after which an observer is left standing in the 'warm sector'.

Clouds formed in the 'warm sector' of a depression

In the warm sector (Figs. 4.12 and 4.13), the amount and type of cloud seen depends
on how far the observer is from the centre of the low pressure. Close to it there is
likely to be unbroken stratus or nimbostratus with rain. Further away, if the air is
unstable, cumulus clouds can develop and, if stable, stratiform cloud may obscure
the approaching cold front.

Clouds formed at the cold front and beyond

Because the slope of the cold front is steeper than that of the warm front
(Fig. 4.13), the succession of clouds and their duration is speeded up. Otherwise
they are the same types of cloud that form at warm fronts, but in the reverse order –
stratus, nimbostratus, altostratus and cirrus. As the cold front passes away to
the northeast, the observer is now standing in the polar air. This cooler unstable air allows extensive convective clouds to form, which are very obvious features

in satellite images, peppered over the dark, otherwise cloud-free areas between (Fig. 4.14). But it is often the case that another depression is forming to the south-west and the whole sequence may be repeated several times, over a period of a few days.

Clouds formed by uplift over mountains

If the wind encounters higher ground – hills or mountains or an island in the sea – the air is forced to rise and if the dew point is reached, clouds will form. Known as *orographic clouds*, these are the same types of cloud as we have already met, the genus formed by the mountain depending mostly on the stability of the air. If stable, stratus or altostratus will form, possibly as a cap cloud sitting over the mountain peak. If unstable, cumulus will form, ranging from the small humilis to large cumulonimbus. As the air passes beyond the mountain, it drops into the lee of the obstruction and so is compressed and in consequence warmed, the effect being enhanced due to the warming of the air induced by the condensation, and so clouds tend to evaporate. This can lead to higher temperatures on the lee of the mountain compared with the temperature on the windward side at the same height. This is known as the *föhn* effect. While the clouds on mountain tops may look stationary, the wind is actually blowing through them, the clouds continuously forming upwind and evaporating in the mountain's lee. Charles Darwin (1839) wrote a note on this in his diary of the voyage of the *Beagle* while observing clouds on the Corcovado Mountain in Rio de Janeiro, saying

The sun was setting, and a gentle southerly breeze, striking against the southern side of the rock, mingled its current with the colder air above: the vapour was thus condensed: but as the light wreaths of cloud passed over the ridge, and came within the influence of the warmer atmosphere of the northern sloping bank, they were immediately re-dissolved.

If there are several discrete humid layers one above the other, a pile of lenticular clouds may form vertically one above the other (Fig. 4.20). These are known as *pile d'assiettes* or 'pile of plates' clouds. While the general shape of lenticular clouds is lens-shaped, the clouds forming them can be any of the varieties seen in normal clouds.

An unusual type of orographic cloud is the *banner*, or *smoking mountain cloud* (Fig. 4.21), caused by the wind creating a reduced pressure on the lee side of the peak thereby producing eddies that draw air upwards. Again, the increase in height causes the air to expand and cool and the water vapour to condense. The most famous examples are on the Matterhorn and Everest.

Orographic cumulonimbus clouds can produce very intense rain and cause flash floods in mountainous areas. I encountered such rain, some of the heaviest I have seen anywhere in the world, even compared with the tropics, in the Snowdonia

Figure 4.20 This *pile d'assiettes* or 'pile of plates' cloud was seen over the Antarctic Peninsula from the British Antarctic Survey ship *Bransfield* en route to the Faraday base.

mountains in Wales one summer while on a fishing trip with one of my grandsons. The intense rainfall was brief but very impressive indeed. We had to stop driving for ten minutes. A mile away it was dry and we fished for trout in the sun.

Very high clouds

Although the majority of clouds are within the troposphere, two types occur that are much higher. Although neither produces precipitation, they are worth a brief mention.

Figure 4.21 On the same voyage as in the previous figure, these banner clouds were forming on isolated mountain peaks along the Antarctic Peninsula.

Noctilucent clouds

Meaning 'shining at night' these are the highest of all clouds, forming at the top of the mesosphere at around 80 km altitude (Fig. 3.4). They are visible for about a month before and after midsummer at latitudes greater than around 45° north and south, lit by the sun while the ground is in darkness. At midnight they are bluish-grey while earlier or later they may be yellow, but being very thin they may not be visible directly overhead, only at an oblique angle from a distance, looking like cirrus or cirrostratus. The particles are of ice, but the nucleation process (see next chapter) is only partly understood, the particles possibly forming on meteoritic dust or on ions produced by cosmic rays.

Nacreous clouds

These 'mother of pearl' clouds form in the mid stratosphere (Fig. 3.4) and can occasionally be seen at latitudes greater than 50° north and south just before sunrise or after sunset. Their colours are very striking, being pastel-shaded, caused by iridescence similar to that sometimes seen at the edges of thin cloud close to the sun (and so not often noticed). The clouds belong to a group known meteorologically as 'polar stratospheric', but occurring only rarely, perhaps once a decade, they arouse comment when they do.

Measuring clouds

Observing cloud cover and type

There are as yet no automatic methods of measuring cloud cover and cloud type, which still have to be estimated by eye along the lines described at the start of the chapter. Developments are under way, however, to observe clouds from the ground remotely using CCTV (Hatton *et al.* 1998, Rowbottom *et al.* 1998). Thermal IR cameras sensing the 8–14 μm spectral range are the most useful since they are able to operate day and night using the natural radiation emitted by the clouds. Because clouds at different heights radiate according to their temperature, the cameras are able to locate the edges of the clouds more easily, unlike cameras working in the visible spectrum, and unlike observers' eyes. Infrared cameras can also penetrate haze.

At present the major limitation to the general use in meteorology of CCTV is the high cost of the IR cameras (£20 000) since they are not widely used generally and so have a small market; however, this may change. So far, cameras have been investigated for the remote *manual* observation of clouds, but the next step is to develop techniques for the *automatic* identification of cloud cover and type. This, however, is some time off because of the subtle variety of clouds, one type merging onto another, and the inevitable complexity of the algorithms required to model them. It is difficult to see how cloud species and varieties could be adequately distinguished automatically. But with the increasing move towards automating meteorological observations, the remote and automatic identification of clouds is becoming a pressing matter. As often it is partly a matter of funding.

Estimating cloud height by eye

While techniques to measure cloud height automatically have been available for some time (see below), at stations without measuring equipment height must still be estimated by eye. In mountainous areas it is possible to judge cloud height if the base is lower than the tops of the hills by reference to topographical features of known height. At all other locations, a table must be used that gives heights for the 10 genera of clouds, different tables being required for the three latitudinal regions. Table 4.6 is for temperate regions.

Measuring cloud height manually by balloon

The height of the cloud base can sometimes be estimated by measuring the time taken for a small rubber balloon, inflated with hydrogen or helium, to enter the misty layer before finally disappearing into the cloud. The rate of ascent is determined by the free lift of the balloon, which can be adjusted by controlling the degree of inflation. Due to eddies near the ground, the balloon may not start to rise immediately and allowance must be made for this in the timing. In addition, the rate of ascent can be somewhat uncertain up to 2000 feet (600 m), and rain reduces the rate of climb. Vertical air currents and the shape of the balloon also affect ascent rate. Without binoculars, the method is limited to about 3000 feet (900 m) unless the wind is very light, because the balloon will disappear from view before entering the cloud. At night a light can be attached to the balloon. The method is clearly not particularly accurate and is not applicable to high clouds.

Measuring cloud height manually by searchlight

At night it is possible to make fairly precise manual measurements of cloud height using a searchlight, the angle of elevation of a patch of light produced by a vertically pointing searchlight being measured some distance away using an *alidad* (Fig. 4.22). The height of the cloud base is given by $h = L \tan E$, where L is the distance between the light and the alidad and E is the angle of elevation. The best separation distance is from 650 to 2000 feet (200–600 m). The largest error originates when measuring the angle of the light spot, $1°$ producing an error of 17 feet (5 m) when the cloud base is 1000 feet, and 450 feet (140 m) when it is 5000 feet, hence the importance of correct alidad installation and care in reading it.

Measuring cloud height automatically by laser

To obtain a continuous record of cloud height, to make measurements at an unattended site, or to get readings both day and night, a *rotating-beam ceilometer* was

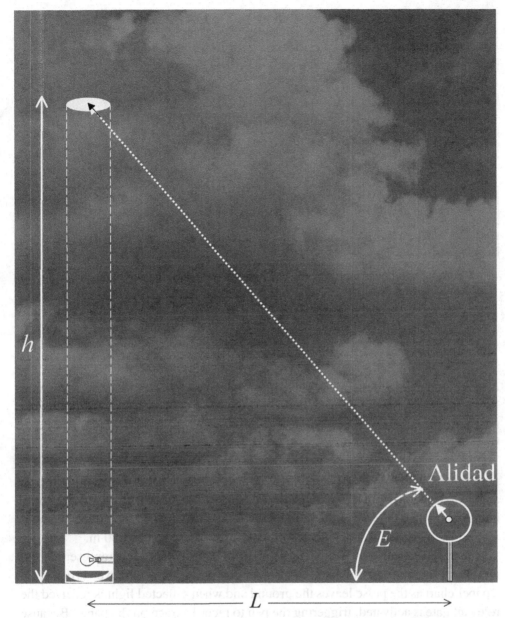

Figure 4.22 At night it is possible to measure cloud height using a searchlight, the height of the cloud base being $h = L \tan E$.

developed during the 1960s using a searchlight, but this will not be described as it has now been largely replaced by laser methods.

By measuring the flight time of a light pulse from a gallium arsenide semiconductor laser, transmitted to the cloud base and reflected back to a receiver, the height

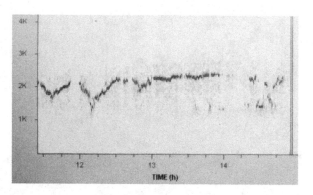

Figure 4.23 Although until recently cloud height was measured automatically using an automated searchlight system, today a laser instrument is used instead (left). To the right is a display on the computer screen of the reflected signals showing time along the *x*-axis and height of the cloud base in kilometres up the *y*-axis.

of the cloud can be determined. In a typical design, the laser is at the focus of a Newtonian reflecting telescope with a mirror of about 200 mm diameter, producing a beam of 8 minutes of arc. The laser emits around 40-watt, 100-ns pulses at a wavelength of 900 nm, with a repetition rate of 1000 Hz (Fig. 4.23).

The receiver has the same Newtonian construction, but in place of the laser a photodiode, with a narrow-band optical filter (to remove natural radiation) and an angle of view of about 15 minutes of arc, senses the returned pulse. With these angles and when the transmitter and receiver are fixed side by side, their fields of view start to overlap at a height of 5 m with full convergence at 300 m.

The received pulse is passed to a series of time gates, each representing an increase of flight time equivalent to 15 m height. A pen starts to move vertically up a paper chart as the pulse leaves the ground and when reflected light is received the relevant gate is activated, triggering the pen to record a trace on the paper. Because of the low speed of the mechanical pen compared with the speed of light, the laser is fired many times during one pen scan. Alternatively the height can be displayed on a computer screen as in Figure 4.23, the range covered being from 100 feet (30 m) to 5000 feet (1500 m).

We can now identify, name and measure the many different types of cloud, but what turns water vapour into cloud droplets and the droplets into rain, snow and hail? The next chapter explores this question. But the foundations had already been laid in the nineteenth century.

References

Bader, M. J., Forbes, G. S., Grant, J. R., Lilley, R. B. E. and Waters, A. J. (1995). *Images in Weather Forecasting*. Cambridge: Cambridge University Press.

Bjerknes, J. and Solberg, H. (1922). Life cycle of cyclones and the polar front theory of atmospheric circulation. *Geofysiske Publikationer*, **3**, 3–18.

Buontempo, C., Flentje, H. and Kiemie, C. (2006). In the eye of a tropical cyclone. *Weather*, **61**, 47–50.

Collier, C. G. (1996). *Applications of Weather Radar Systems: a Guide to Uses of Radar Data in Meteorology and Hydrology*, 2nd edn. Chichester: John Wiley/Praxis.

Collier, C. G. (2003). On the formation of stratiform and convective cloud. *Weather*, **58**, 62–69.

Darwin, C. (1839). *The Voyage of the Beagle*. London: Henry Colburn. Republished Harmondsworth: Penguin, 1989.

Dunlop, S. (2003). *The Weather Identification Handbook*. Guilford, CT: Lyons Press.

FitzRoy, R. (1863). *The Weather Book*. London: Longman, Green, Longman, Roberts & Green.

Hamblyn, R. (2001). *The Invention of Clouds*. London: Picador.

Hatton, D., Jones, D. W. and Rowbottom, C. M. (1998). *Technical Experiences of Using a Video Camera to Make Remote Weather Observations*. WMO Instruments and Observing Methods, TECO-98, May 1998.

Houze, R. A., Jr. (1993). *Cloud Dynamics*. San Diego, CA: Academic Press.

Howard, L. (1804). *On the Modifications of Clouds, and on the Principles of their Production, Suspension, and Destruction: being the substance of an essay read before the Askesian Society in the session 1802–3*. London: J. Taylor.

Jonas, P. R. (1994). Back to basics: why do clouds form. *Weather*, **49**, 176–180.

Mason, B. J. (1957). *The Physics of Clouds*, 2nd edn. Oxford: Clarendon Press.

McGinnigle, J. B., Young, M. V. and Bader, M. J. (1988). The development of instant occlusions in the North Atlantic. *Meteorological Magazine*, **117**, 325–341.

Met Office (1982). *Observer's Handbook*. London: HMSO.

Pearce, R. (2005). Why must hurricanes have eyes? *Weather*, **60**, 19–24.

Rowbottom, C. M., Hatton, D. B. and Jones, D. W. (1998). *Operational Experiences in the Use of Video Camera Images to Augment Present Weather Observations*. WMO Instruments and Observing Methods, TECO-98, May 1998.

Scorer, R. S. (1972). *Clouds of the World*. Newton Abbot: David and Charles.

Scorer, R. S. (1994). *Cloud Investigations by Satellite*. Chichester: John Wiley/Praxis.

Scorer, R. S. and Verkask, A. (1989). *Spacious Skies*. Newton Abbot: David and Charles.

Weather (2003). Special issue. Luke Howard and clouds: a bicentennial celebration. *Weather*, **58**, 49–100.

WMO (1975). *International Cloud Atlas. Vol. I: Manual on the Observation of Clouds and Other Meteors*. WMO 407. Geneva: World Meteorological Organization. Reprinted 1995.

WMO (1987). *International Cloud Atlas. Vol. II (plates)*. WMO 407. Geneva: World Meteorological Organization.

5

Cloud droplets, ice particles and precipitation

By the start of the twentieth century a simple basic understanding of how cloud droplets form had been reached. It is useful to review these findings before going forward to fill in the details learnt during the last century. Recall, for example, the experimental work of Guericke in the seventeenth century with his air pump and cloud chamber in which a few large droplets might form one day, a large number of smaller drops on another. He wrongly concluded that two different processes were in action. A century later Kratzenstein repeated Guericke's experiments but was confused by the colours he saw and wrongly concluded that the drops were bubbles (or 'vesicles'). In the nineteenth century, Waller collected cloud droplets on spider webs and concluded that they were drops, not bubbles. Then in 1875 the Frenchman, Coulier, while doing experiments with a cloud chamber, deduced correctly that droplets only form if there is 'dust' in the air. This finding was ignored at the time, but soon afterwards John Aitken, in Scotland, made a similar discovery. Aitken acknowledged Coulier's work when his attention was drawn to it, but went on to conclude that the fewer the 'dust' particles, the larger the drop sizes and the smaller their number, suggesting that the available water vapour was shared amongst the particles. He also concluded that most probably dried sea spray in the form of fine powder was the main source of nuclei, other possible causes being volcanic and meteoric dust, along with smoke from combustion. Tests on a variety of finely divided substances showed that they differed greatly in their nucleating power, hygroscopic nuclei being the most active, which includes both salt and the hygroscopic products of combustion. So what more have we learnt since Aitken came to these conclusions at the end of the nineteenth century?

While the basic processes leading to the formation of clouds had been established by the early twentieth century, details of how the droplets actually form were only clarified during the second half of the century, through a study of the microphysics of clouds. This is what we look at in this chapter.

Droplet formation in warm clouds

By the WMO definition, clouds are 'a visible aggregate of minute particles of water or ice, or both, in free air'. By convention, drops in the range 0.1 to 0.25 mm radius are known as *drizzle* while drops larger than this are *rain*. Smaller drops are cloud droplets.

Nucleation

Just as air molecules continually collide, so too do water vapour molecules, and when so doing they can combine to form a liquid-phase drop. This process is termed homogeneous nucleation (to distinguish it from heterogeneous nucleation, in which the water vapour molecules collect on a foreign body). The net energy needed to accomplish nucleation (Houze 1993) is:

$$\Delta E = 4\pi R^2 \sigma_{vl} - 4/3\pi R^3 n_l(\mu_v - \mu_l) \tag{5.1}$$

where σ_{vl} is the work required to create a unit area of the surface of the vapour–liquid interface round the drop – or surface energy or tension

n_l is the number of water molecules per unit volume of drop

μ_v is the Gibbs free energy of a single water vapour molecule

μ_l is the Gibbs free energy of a single liquid water molecule

If the energy needed is more than the change of Gibbs free energy ($\Delta E > 0$) the drop formed (by chance union of molecules) cannot be maintained and evaporates. If, however, the required energy is less than that available ($\Delta E < 0$) the drop survives and is said to have nucleated.

It can be shown that the difference in free energy is:

$$\mu_v - \mu_l = k_B T \log_e(e/e_s) \tag{5.2}$$

where e is the vapour pressure

e_s is the saturation vapour pressure

k_B is Boltzmann's constant

T is the temperature

By substitution in (5.1), and rearranging the terms, an expression can be derived for the critical radius R_c of a drop in which the two terms in (5.1) are in equilibrium:

$$R_c = 2\sigma_{vl}/n_l k_B T \log_e(e/e_s) \tag{5.3}$$

This is known as Kelvin's formula, and from it can be seen that the critical radius, R_c, is very dependent on the relative humidity ($e/e_s \times 100\%$). When the air is just saturated ($e/e_s = 1$), $R_c \to \infty$, so for a drop to form the air must be supersaturated ($e/e_s > 1$). The higher the supersaturation, the smaller the drop that can survive.

The value of R_c is also dependent on the temperature, but at normal atmospheric temperatures its effects are weak.

So the rate of nucleation of drops reaching the critical size depends mostly on the degree of supersaturation. But to nucleate homogeneously in any significant numbers the air must be supersaturated by 300–400%, and as the atmosphere rarely exceeds 1% supersaturation, the homogeneous nucleation of cloud droplets plays no role in forming natural clouds.

However, as soon as nuclei are present in the atmosphere as aerosol particles, water vapour will also collide with them and can collect on their surface. This is known as heterogeneous nucleation and is the process by which normal clouds form. If the surface tension between the water and the surface of the nucleus is low enough, the nucleus is said to be wettable and the water can form a spherical cap on its surface. Such a particle is known as a *cloud condensation nucleus* (CCN). The water, at this stage, does not completely surround the nucleus; it simply forms a domed cap on part of it.

If the particle is insoluble in water, the microphysics is the same as in the previous case of homogeneous nucleation and Equation (5.3) still applies, although now the radius is not that of a fully rounded drop but the *radius of curvature* of the embryonic drop on the nucleus. As this will be much larger than the drop that would form in the absence of a nucleus from the same number of vapour molecules, there is a greater chance that it will survive. If sufficient water molecules collide with the particle they may form a complete film of water all round the nucleus, forming a drop that is much larger than it would be if there was no nucleus at its centre. For this reason, the larger the nucleus the better the chances of the drop surviving.

But if the nucleus is also soluble in water, its effectiveness is increased because the saturation vapour pressure e_s is lower over a solution than over pure water, thereby increasing e/e_s and making the critical radius smaller. For this reason, salt nuclei are particularly efficient at drop formation. The natural atmosphere is not clean and there are more than enough wettable particles available to form clouds; their availability is not a limiting factor. But from the above, it is evident that the first droplets will form on the larger and most soluble particles. The size and composition of aerosols consequently has a great effect on the distribution of drop sizes within a cloud. Having formed, cloud droplets can continue to grow as they encounter more vapour molecules by diffusion, and these condense onto them. The reverse process can also occur as vapour evaporates from the droplets.

Coalescence into raindrops

Having formed viable drops that survive, the drops can continue to grow by combining with other drops during collision as they fall. The *collection efficiency* is the

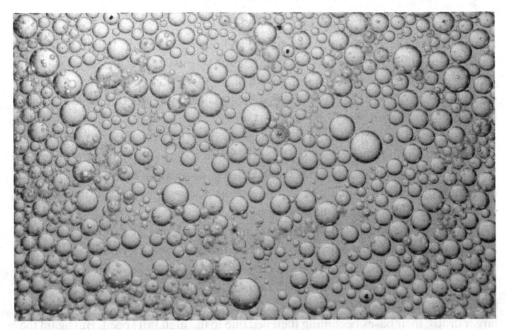

Figure 5.1 A microphotograph of raindrops collected in a dish of oil during a heavy thunderstorm, showing the range of drop sizes. The largest are 5 mm diameter (the largest size a raindrop can reach without breaking up), the smallest is less than 0.5 mm. There can, therefore, be a wide range of drop sizes in the course of one rain event.

efficiency with which a drop intercepts and joins with the drops it is overtaking, this being the product of *collision efficiency* and a *coalescence efficiency*. The collision efficiency is determined by the airflow around the drops as they fall, smaller particles sometimes being carried out of the path of the larger, faster drops, or sometimes drops not directly in line may be drawn in. The *coalescence efficiency* reflects the fact that just because two drops might collide, they may not actually coalesce permanently, perhaps breaking away after uniting or perhaps simply bouncing off each other. However, normally coalescence is an efficient process and it is usual to assume that it is 100% (although the process is still not fully understood in detail). The collection efficiency is thus reduced to the collision efficiency.

Coalescence is not a continuous process but goes in steps as drops unite. But some drops undergo more collisions than others and thus a distribution of drop sizes develops (Fig. 5.1). As time passes, a large part of the available water droplets becomes concentrated in the larger drops. While initially the cloud droplets are far more numerous than raindrops, after a relative short time (tens of minutes) the raindrops contain a large part of the overall liquid water. The process of coalescence can, therefore, quickly convert cloud water to rainwater.

Fall speed of drops

Drizzle and raindrops are subject to gravity and this leads to their fallout as precipitation. This effect is partly offset by the frictional resistance of the air, and drops soon reach their *terminal fall speed V*, which is a function of their radius *R*. *V* is small until *R* reaches about 0.1 mm, this being taken to be the threshold between cloud droplets and precipitation.

The terminal fall speed of drizzle increase linearly with radius from around 1 m s^{-1} at 0.1 mm radius to 3.5 m s^{-1} at 0.25 mm radius (Beard and Prupparcher 1969). But the speed also depends on air density, decreasing as the drops fall and encounter denser air, so that by the time a drop reaches ground level its speed is about 20% less than at an altitude of 18 000 feet.

For rain, the terminal fall speed is not linear, increasing at a slower rate with increasing size, such that a drop of 0.5 mm radius has a velocity of about 4 m s^{-1} while at 1.0 mm radius it has reached 6 m s^{-1}, the increase in speed levelling off rapidly thereafter, so that drops of 2.5 mm radius fall at 9 m s^{-1}. The speed does not increase further beyond about 5 mm diameter due to the change of shape of the larger drops, their bases becoming flattened due to the high fall speed. But again the fall speed depends on altitude, the difference between ground level and 18 000 feet being around 35% for the larger drops. Of course if the air happens to be rising, such as in a cumulonimbus cloud, at say 9 m s^{-1}, even the largest drops will remain suspended and not fall relative to the ground.

After reaching a certain size, however, raindrops become unstable and break up into smaller drops. This has been studied in laboratory tests and empirical relationships have been obtained showing that drops with a diameter of 3.5 mm or less rarely divide. The rate of break-up increases exponentially for drops larger than this and the probability of drops much greater than 4 mm fragmenting is 100%.

Microphysics of cold clouds

Nucleation of ice particles

In principle, ice particles in clouds can be nucleated directly from the vapour phase in the same way as a drop, and the critical size for its formation is given by an equation similar to (5.3) above. In this case, however, the critical size is strongly dependent on both temperature and humidity, and nucleation only occurs at temperatures below −65 °C and at supersaturations of around 1000%. Such conditions do not occur in the free atmosphere and for this reason homogeneous nucleation of ice directly from the vapour phase never occurs in natural clouds.

From observation of clouds, it is known that ice crystals form between 0 °C and −40 °C. Since it is known that homogeneous nucleation does not occur at these temperatures, the crystals must be forming by a heterogeneous process. Just as heterogeneous nucleation needs aerosol particles on which to form in the case of liquid drops, the same is also true for ice. But ice does not form so easily on the particles normally found in the atmosphere because the molecules of the solid phase of water form in an ordered crystal lattice. For ice to form on a foreign substance, the foreign substance must have a lattice structure similar to ice. In the case where the lattices are different, either the ice lattice has to deform or it must have dislocations at the ice–substrate interface. Both of these reduce the nucleating efficiency by increasing the surface tension between ice and substrate or by raising the free energy of the ice molecules. Ice itself is, therefore, the best nucleating material and whenever a supercooled drop at any temperature at, or below, 0 °C meets an ice surface it freezes immediately. Natural materials with a crystalline structure most akin to ice appear to be certain clays that occur in many types of soil and bacteria that live on decaying leaves, in particular *Pseudomonas syringae* and *Erwinia herbicola*. These can cause nucleation at a temperature as high as −4 °C but they do not occur in high concentrations in the atmosphere, and so most ice-particle nucleation occurs at much lower temperatures.

Particles on which ice crystals can nucleate are called *ice nuclei*, and these will initiate freezing of a supercooled drop when the temperature falls low enough to 'activate' the nucleus. If the nucleus on which the drop initially formed also happens to be an ice nucleus, the process is known as *condensation nucleation*. Drops can also be induced to freeze if an ice nucleus in the air comes into contact with an already-formed supercooled drop which nucleated on a non-ice nucleus, such as a salt nucleus, this being termed *contact nucleation*. Finally the ice can form on an ice nucleus directly from the vapour phase, this being called *deposition nucleation*.

Measurements in expansion chambers have enabled an empirical relationship to be derived giving the average number, N_l, of ice nuclei per litre of air:

$$\text{Log}_e \, N_l = a_l(253 \, \text{K} - T) \tag{5.4}$$

where a_l is a constant that varies with location, but generally falls between 0.3 and 0.8, and T is the air temperature.

From this equation it can be calculated that at −20 °C the concentration is only about one nucleus per litre, rising exponentially by a factor of around ten for every 4 °C lowering of temperature.

However, in natural clouds the concentration of ice particles at any particular temperature has often been found to be much greater (by one or two orders of magnitude) than is predicted by Equation (5.4), although the effect is not uniform in space and time, large numbers of ice particles occurring preferentially in the older part of a cloud. The higher numbers start near the top of the cloud and can increase a thousand-fold in ten minutes. How the number of ice particles increases so greatly compared with the number of nuclei predicted by Equation (5.4) is not known, but several possibilities have been proposed (Houze 1993):

Delicate ice crystals may shatter when they collide.

When large supercooled droplets collide with an ice surface between −3 and −8 °C, small ice splinters result.

Certain aerosols may cause nucleation to occur at higher temperatures through *contact* nucleation than they would through other forms of nucleation.

Condensation and deposition nucleation may be much increased when supercooling rises to 1%. (Equation (5.4) was derived from cloud chamber experiments with the air just under saturation. A few percentage points are probably crucial.)

In other words, because the cloud-chamber laboratory tests do not simulate very well the much more complex reality of natural clouds, Equation (5.4) is but a rough guide.

Growth of snow particles by deposition of water vapour

Following initial nucleation, the embryonic ice crystals can grow as water vapour molecules come into contact with them through the process of *deposition* (the ice-phase equivalent to condensation). Deposition is, as noted above, also one of the processes by which initial nucleation can occur. The reverse process can also happen when vapour leaves the surface and diffuses away, a process known as *sublimation* (the ice-phase equivalent to evaporation).

The shape the crystals take as they build by deposition depends on the temperature and degree of supersaturation. The fundamental shape is always the familiar hexagon, but this can be in the form of a thin flat plate or an elongated prism-like column. There can occasionally be twelve-sided crystals. If the temperature changes while they are forming, a composite crystal results with the characteristics of several forms. An increase in the level of supersaturation increases the surface-to-volume ratio of the crystal, resulting in more complex dendrite (tree-shaped) crystals – the conventional image of a snowflake. Variations are large and a table of 80 shapes was produced by Magono and Lee (1966) (this can be seen, along with other illustrations, at http://www.its.caltech.edu/~atomic/snowcrystals). Figure 5.2 illustrates the more common forms of snow crystal, while Table 5.1 shows the temperatures at which they form. A new book of photographs of snow crystals (Libbrecht and Rasmussen 2004) illustrates many of the types mentioned here.

Growth of snow by aggregation

If snow particles that have formed by deposition subsequently adhere to each other, the process is known as *aggregation*. This does not usually occur much below

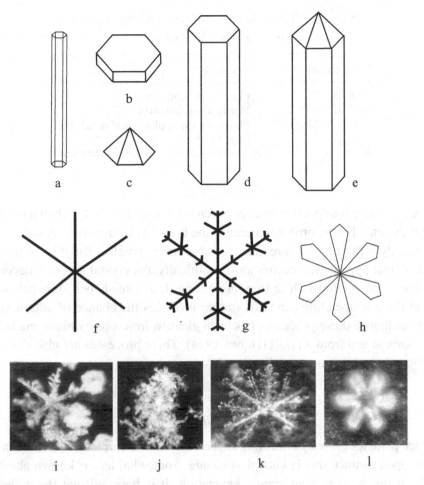

Figure 5.2 The basic shapes of snow crystals:

(a) needles, which can also form as parallel bundles or in a crossed combination; some are hollow (sheaths)
(b) hexagonal plates can occur in a variety of thicknesses
(c) pyramids
(d) columns, which can be solid or hollow
(e) bullets, which can be solid or hollow
(f) stellar crystals, which can have plates on the ends of the arms
(g) dendritic (tree-like) crystals – the traditional snowflake; there is a large range of possible shapes (Libbrecht and Rasmussen 2004)
(h) sectored plates
(i) dendritic crystals with plates on two of the arms, partially rimed
(j) spatial dendrite – an aggregation of many individual crystals
(k) twelve-armed dendrite, imperfectly formed
(l) sectored plate (as h) with rime

The photographs are my own, taken with very simple equipment and so not up to the standard of specialists in the art. However, they illustrate the shapes well enough.

Table 5.1 *Temperature range over which various snow crystal shapes form.*

Temperature (°C)	Crystal type
0 to −5	Plates and dendrites
−5 to −10	Needles and columns
−10 to −22	Plates, sectored plates and dendrites
−22 to −35	Plates and columns

−20 °C, but there is a specific band, between −10 °C and −16 °C, when it becomes more prevalent. This is probably because the branches of dendrite crystals, which grow mostly in this temperature range, can become entangled. But it is from −5 °C to +2 °C that aggregation occurs most prolifically, the crystal surfaces becoming adhesive at temperatures close to or at melting. Dendrite-shaped snowflakes also form at these temperatures and this further increases the chance of adhesion. An understanding of these processes has been gleaned from observations made both on the ground and from aircraft (Hobbs 1974). These processes are also discussed by Libbrecht and Rasmussen (2004).

Growth of hail by riming

If the ice particles grow by collecting supercooled cloud droplets that freeze immediately upon contact, this is known as *riming*. Somewhat less is known about the details of this process than about aggregation. It is believed that the collection efficiency of rime is high, probably near unity. If riming is light to moderate, the crystal's original shape is still recognisable, but when heavy the original shape of the crystal is completely hidden, the resulting particle being known as *graupel* or *soft hail*, taking the shape of lumps or cones.

When riming is extreme hailstones are the result, and these can vary in size from a few millimetres to ten or more centimetres. Hail results when graupel or frozen raindrops collect further supercooled cloud droplets. The latent heat of fusion of ice released when the water freezes warms the hailstones but so long as the surface of the hail remains below 0 °C it remains dry and its development is known as *dry growth*. However, if the hailstone remains in a supercooled cloud long enough the rate of riming can be so fast that the loss of heat is less than its generation and the surface temperature can rise above freezing point. When this occurs, the supercooled droplets no longer freeze instantly upon contact and some may not adhere, although most of the water turns into a water/ice mesh called *spongy hail*;

this is known as *wet growth*. Hail may pass through a combination of dry and wet growth stages as it falls through air at different temperatures and through droplets at different levels of supercooling. When this occurs, hailstones, cut through to reveal their internal structure, show evidence of a layered formation. The Greeks noticed this, but could not explain it.

While working on the Cairngorm mountains in Scotland in connection with the development of instruments for cold regions, we frequently experienced the effects of riming throughout the winter. At 4000 feet (1220 m) in a maritime climate and at this high latitude, clouds were often supercooled and the windspeed high. Riming could be so severe that instruments became completely encased, inoperable and often damaged. This is the same process that produces hailstones and it could be experienced first-hand without the need for aircraft. Figure 5.3 illustrates a marker post encased in rime.

Fall speed of ice particles

During the twentieth century a number of people made careful measurements of the terminal fall speeds of snow and hail, amongst these being Nakaya and Terada (1935), Magono and Lee (1966), Hobbs (1974) and Hobbs and Rango (1985). Figure 5.4 is my combined summary of what they found, all of this being presented in more detail in Houze (1993). In the case of hail, Pruppacher and Klett (1978) derived the following simple empirical equation for its fall velocities at a pressure of 800 hPa and a temperature of 0 °C:

$$V = 9D^{0.8} \tag{5.5}$$

where V is the velocity in m s^{-1} and D is the diameter of the hail in centimetres. This was derived for hailstones ranging from 1 mm to 8 cm. Using this equation it can be calculated that fall speeds range from 10 to 50 m s^{-1}. Houze (1993) observes that these large values imply that updraughts of similar velocity must exist in the cloud to support the hailstones long enough for them to grow. For this reason, hail is only found in very intense thunderstorms.

Melting

As the ice particles fall, they pass through air at different temperatures and relative humidities, possibly through air containing cloud droplets (supercooled or not) and raindrops. The rate of melting depends on the extent and direction of heat transfer between the ice particles and the air and water droplets. There can also be a gain or loss of heat through condensation onto the particles or evaporation from them.

Figure 5.3 Rime ice formed on a post near the summit of Cairn Gorm in the Scottish Highlands. Rime builds into the wind as the supercooled cloud droplets impact the pole, freezing immediately.

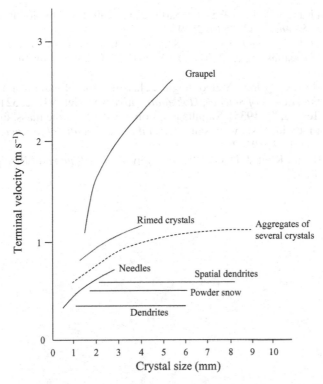

Figure 5.4 Whereas raindrops differ only in their diameter, snowflakes can take on many different forms and so their fall velocity varies considerably depending on their shape. This diagram is an adaptation and combination of several figures in Houze (1993) and shows the terminal velocity of the various types of snow crystal and graupel. The fall velocity of hailstones is an order of magnitude greater than for graupel.

This completes our examination of what precipitation is, how it forms and how our views on this have evolved since the days of the Greeks. Having arrived here after that long journey of 2500 years, I cannot help but reflect 'Why did it take so long?'

Before moving on to see how precipitation is measured, we will detour briefly to take a look at lightning.

References

Beard, K. V. and Pruppacher, H. R. (1969). A determination of the terminal velocity and drag of small water drops by means of a wind tunnel. *Journal of the Atmospheric Sciences*, **26**, 1066–1072.
Hobbs, P. V. (1974). *Ice Physics*. Oxford: Clarendon Press.

Hobbs, P. V. and Rango, A. L. (1985). Ice particle concentrations in clouds. *Journal of the Atmospheric Sciences*, **42**, 2523–2549.

Houze, R. A., Jr. (1993) *Cloud Dynamics*. San Diego, CA: Academic Press.

Libbrecht, K. and Rasmussen, P. (2004). *The Snowflake*. Grantown-on-Spey: Colin Baxter Photography.

Magono, C. and Lee, C. (1966). Meteorological classification of natural snow crystals. *Journal of the Faculty of Science, Hokkaido University*, Ser. VII, **2**, 321–325.

Nakaya, U. and Terada, T. (1935). Simultaneous observations of the mass, falling velocity, and form of individual snow crystals. *Journal of the Faculty of Science, Hokkaido University*, Ser. II, **1**, 191–201.

Pruppacher, H. R. and Klett, J. D. (1978). *Microphysics of Clouds and Precipitation*. Dordrecht: Reide.

6

Lightning

Lightning was probably the cause of the first fires seen and used by humans, but before Benjamin Franklin it was just a mysterious and terrifying natural phenomenon, although others had been debating the matter before this. In 1752, in addition to all his political activity, Franklin found time, aged 46, to perform his famous kite experiment, in which he established that lightning was due to static electricity, akin to laboratory-produced electrical discharges. This work was published by the Royal Society in London and four years later he was elected a Member. In 1772 he was also elected to the French Academy of Sciences. It is interesting to reflect that despite all of his other work, his world fame resulted from this single experiment and his subsequent invention of the lightning rod.

The clouds that produce lightning

As we saw in Chapter 4, single-, multi- and super-cell cumulonimbus clouds involve powerful convective updraughts producing the highest and most violent of all clouds, generating hail, tornadoes and hurricanes. In addition they also produce lightning. By what process are the clouds charged? This is still undecided and is the subject of ongoing research, but nevertheless a great deal is now known.

The charging process

During a thunderstorm, the upper regions of the cloud become positively charged, with a negative charge near the centre of the cloud at around the $-15\ ^\circ$C level in a pancake-shaped layer about 1 km thick, extending over the width of the cloud (Fig. 6.1). This in turn attracts a positive charge on the ground below, which travels along beneath the cloud as it moves. Pockets of different polarity can also be generated at different levels within the cloud, and there is evidence that there can also be a small positive charge near the base of the cloud.

119

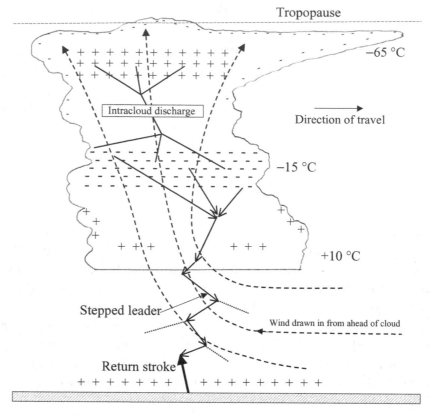

Figure 6.1 The charging process within a cloud has not yet been fully explained, although the charge distribution within the cloud is now well understood. What initiates a strike is also not fully explained yet, although the mechanism of a lightning strike is better researched and documented. See text for details.

There is still no certain theory as to the mechanism of charge separation, but in many storms worldwide the negative region has been found to occur between temperatures of −10 and −25 °C, centred on the −15 °C level, suggesting interactions involving the ice phase. There is now a generally agreed view emerging that the charge separation results from descending graupel particles colliding with small ice particles, the polarity of the charge depending on the air temperature and liquid water content. Below about −15 °C (from −10 to −25 °C) the graupel becomes negatively charged. In warmer air the reverse occurs.

The view currently favoured by the Royal Meteorological Society, in its 'Scientific statement on lightning' (RMS 1994), is that during collisions between ice crystals and small hail pellets, in the presence of supercooled water droplets, the pellets becoming negatively charged while the ice crystals bounce off with a positive charge; this has been demonstrated to occur in laboratory tests. The updraught then carries the crystals aloft, where they produce the upper positive charge, while the pellets are heavy enough to fall against the updraught, forming a negative charge lower down (Fig. 6.1). At warmer temperatures, the charging process reverses, giving the lower region a positive charge. The negative charge shown as a thin layer around the top edge of the cloud is believed to be produced by negative ions (resulting from cosmic ray activity) being attracted to the upper positive charge.

There is also evidence, albeit limited, that clouds completely above the freezing level can become charged, and plumes from volcanoes and forest fires can also produce lightning, as can sandstorms. So the charging processes are still not fully understood and are being actively investigated (Dudhia 1996, 1997).

Initiation of a discharge

Lightning discharges occur when the difference in potential is sufficient to break down the resistance of the insulating air. But there is a problem since such voltages are not high enough, under laboratory conditions, to cause electrical breakdown. One current view to account for this is that transitory filaments of liquid, resulting from glancing collisions between supercooled raindrops, or protuberances from ice crystals or graupel, initiate the discharge. While this might explain the initiation process, it does not explain how the discharge then spans a distance much greater to the ground than it does in the laboratory. However, it is known that initiation of a strike is also related to the nature of the ground surface and to the characteristics of the lower 1 km of the atmosphere. There is also a possibility that changing windspeed with altitude increases the possibility of flashes (Hardaker 1998). Clearly, further research is needed to understand both the charging and the discharging processes.

One old proposal which is now being reconsidered is the idea that cosmic rays may strike an air molecule and ionise it, producing very high-energy electrons. In the electric field of a thundercloud, such an electron could be accelerated, ionising further air molecules in a chain reaction, the resulting avalanche of electrons ionising the air and allowing charge to flow, a process termed 'runaway breakdown' (Gosline 2005).

Lightning strikes

Although the processes producing the charge and initiating the discharge are not yet wholly understood, what actually happens in a thunderstorm is much better recognised. Discharges can be between different parts of the cloud (intra-cloud), or from the cloud to the ground. There is no difference between the two, except that within the cloud the actual path of the discharge is hidden, illuminating the cloud internally, this being known as *sheet lightning*. Discharges to the ground allow the paths to be seen, as a network of tracks akin to the roots of a tree, and this is known as *fork lightning*. There can also, more rarely, be flashes from cloud to cloud and into the clear air beside or above the cloud, the latter being known as *rocket lightning*. Again there is no clear explanation of these.

Over half the flashes (70–90%) are of the sheet-lightning type, but it is the cloud-to-ground strikes that are of more practical concern since these are the ones that cause damage, although the former are of concern to aerospace activities. A ground strike starts near the base of the cloud, as a faintly luminous discharge, producing a stepped *leader* (Fig. 6.1), which travels downwards in discrete intervals of about a microsecond, travelling at around 61 000 miles per second (98 000 km s^{-1}). When the leader approaches within about 100 m of the ground, a return stroke moves up from the ground (usually from a protruding object), the negative charge on the cloud flowing to ground along the ionised path thus established. It is this return stroke that produces the visible lightning, the brightest part of the path moving upwards at about one-third the speed of light. There is often sufficient charge on the cloud to produce several strokes along the same ionised path in the space of less than a second, producing the familiar flickering effect. The combined series of strokes is referred to as the *flash*.

There are rarer occasions when positive discharges are initiated from the top of the cloud, often from the overhang of the cloud's anvil. Positive discharges, for some reason, are more common in the northern hemisphere in winter. These also tend to be one-stroke flashes. The increased use of lightning detectors (see below) is starting to show, however, that positive strokes are more common than were thought, perhaps as many as one in five. Leaders starting from the ground are rarer, but do sometimes occur from positively charged high buildings or mountains. Negative leaders from the ground are still less common.

Currents flowing along the ionised path range from 3 to 300 kA, typically 10 000–30 000 amps, releasing 100 megawatts per metre of channel. This causes violent and rapid heating of the air to about 30 000 °C, which produces a shockwave that travels faster than the speed of sound and a subsequent sound wave that we hear as thunder. The sound is drawn out into a rumble in part because there are several strokes in rapid succession, and also because the sound is not from a single point source, but from the extended path from ground to cloud, sound arriving at different times from different points along the path. Echoing is also involved in drawing out the length of the sound yet further.

There are around 2000 storms in action at any one time worldwide, with about 100 flashes per second. In the UK there is an average ground strike of about one per square kilometre per year. Worldwide, thunderstorms causes a small negative charge to build up on the earth's surface, with an equal number of positive charges in the atmosphere. This causes a voltage gradient of about 300 kV between the ground and the conducting layers of atmosphere at 60–70 km, decreasing with height due to the increasing conductivity of the air so that across the lowest metre there is a potential difference of about 100 volts. Although each flash involves a very high current, each of the 2000 storms around the world produces an average current, worldwide, of only about 1 amp, as the negative discharges flow to ground. A slow, continuous, reverse, air-to-ground leakage current of 2000 amps brings positive charges to ground, balancing the overall current flow.

Ball lightning, Saint Elmo's fire and sprites

Ball lightning is a rare phenomenon – a glowing sphere, ranging in size from a few centimetres to a metre. It is usually coloured red, orange or yellow with a lifetime of up to a minute. The ball can be stationary or move slowly, horizontally, at a few metres per second, ending either silently or with a small explosion. It can occur in- or out-of-doors. Many theories have been proposed to explain it, but none is yet widely accepted. The Royal Society (2002) recently brought together many unpublished sightings, of which there have been around 10 000 over the past few decades, demonstrating that it is real enough. The report contains seven papers in all and concentrates on the chemical effects of electrical discharges. One current hypothesis is that when lightning strikes the ground it may vaporise silica in the soil, which condenses into a ball of fine dust held together by electric charge, although it is probable that no explanation is yet complete (Muir 2001). Durand and Wilson (2006) analyse around 30 papers on ball lightning but come to no firm conclusion, one difficulty being that the formation of the balls has been so rarely observed that the exact cause(s) are still unknown. The authors also discuss a similar phenomena, 'fireballs', about which even less appears to be known. In the year 1793 there were many reports of ball lightning and fireballs and the possible connection with volcanic air pollution, following the eruption of Laki, a volcano in Iceland, is discussed, although again nothing is conclusive.

Saint Elmo's fire is less problematic and consists of a continuous glowing electrical discharge from sharp objects such as ships' masts, church spires, trees and even hair, caused by atmospheric electricity sometimes accompanied by a crackling sound. I had personal experience of similar phenomena while working on meteorological instruments on the summit of Cairn Gorm in Scotland. Raising a screwdriver above my head to make an adjustment to an instrument, a loud hissing sound came from it. Overhead was a large and dark cloud. It is probable that if it had been night time there would have been a glow from the screwdriver. Saint Elmo's fire is so named after the patron saint of Mediterranean sailors, the discharge being said to be a sign of protection. I felt it had the opposite significance on the mountain.

Sprites are incompletely understood phenomena that occur high above thunder-clouds in the mesosphere at 40–80 km altitude following a lightning strike from the positively charged top of the cloud. They take the form of towers of light last-ing only milliseconds, but of large size, typically 10 km across and up to 30 km high. The most usual colours are red or blue. They are fairly common but were not photographed until 1989 and the mechanism of their production is still a matter of debate. There is some suggestion that they emit radio waves, as do normal lightning flashes, but this is not yet confirmed. Research continues, some of the latest with balloons (Williams 2001). Studying these from the International Space Station or from the Shuttle might throw interesting light on their nature.

Units and terms

The radio waves produced by a lightning flash are known as *atmospherics*, abbre-viated to *sferics* (WMO 1996).

The strength of a flash is defined as the peak value of current of the first return stroke, measured in amperes (A), typically tens of kiloamperes (kA).

The intensity of a storm is defined in terms of:

the flash rate in a given area
the average number of strokes in a flash
the polarity of the flashes

Detecting lightning

As Heinrich Hertz discovered in the 1880s, an electrical spark generates radio waves which can travel great distances, and it is the measurement of radio emissions from lightning discharges that forms the basis of all long-distance detection systems. Lightning emits radio waves over a wide spectrum of frequencies from the extremely low frequency (ELF) (3 to 30 Hz, 100 000 to 10 000 km) to very high frequency (VHF) (30 to 300 MHz or 10 m to 1 m), the longer waves being emitted most strongly by the return stroke of a lightning strike. It is the long waves that are used in most systems to detect the location of distant thunderstorms, since they cover vast distances by way of the ground wave.

Methods of detecting lightning at a distance are of two types – those that measure the *direction of arrival* of the sferic pulse and those that measure their *time of arrival* (TOA) (sometimes known as LPATS, or Lightning Position and Tracking System).

A different design, developed by the UK Met Office using the TOA principle, is named the Arrival Time Difference (ATD) method. A short-range system has also been developed that operates in the VHF band. All of these are examined below. The detection of local lightning is, however, done differently, and is considered first.

Local lightning detection

Local observations are made mostly for utilitarian purposes, at places vulnerable to lightning such as explosive stores, electricity installations, airports and rocket launching sites. Cape Kennedy, for example, is very liable to lightning strikes, and shuttle flights can be cancelled at the last minute due to them (I experienced such a cancellation, much to my disappointment). Since local measurements are not widely used for meteorology or scientifically, however, very little has been reported in the literature regarding the methods of measurement or their performance (Johnson and Janota 1982, WMO 1996). The techniques are thus only briefly mentioned here.

Visual observation

By timing the delay in arrival of the sound of thunder following a flash, a rough measure of distance can be obtained without the need for instruments, every three seconds being equivalent to one kilometre, or five seconds to the mile. But such observations are clearly very limited.

Measuring the static electric field

For automatic and more precise cover, the vertical static electric field in the atmosphere in the vicinity of a cumulonimbus cloud can be measured to give advanced warning of the likelihood of lightning. Since the electrical field decreases rapidly with distance from the cloud, the detection range is limited to about 20 km, although a network of such instruments can give cover for as wide an area as is instrumented.

There are several methods of measuring the vertical field, a common technique being the *electric field mill*, in which the potential gradient in the atmosphere is measured by an electrode fixed above a metal plate about a metre below it at earth potential. In one such design developed in France, a motor-driven rotating shutter alternately exposes and shields the sensing electrode from the field (in what is in effect a Faraday cage), producing a sinusoidal AC signal from the DC field. This minimises measurement drift and noise. A switch on the motor shaft, synchronised with the rotating shutter, provides the means to sense the phase of the AC signal and thus allow the peak level of the sine wave signal to be sampled and stored. The field strength can be computed from this and related to the probability of a flash. In another instrument, the *flash counter*, the rapid *change* of vertical electric field when lightning occurs is measured.

A high, sharp-pointed conductor, connected to the ground, emits a corona discharge when the local static field increases beyond a certain threshold. By measuring the discharge current, the strength of the field can be estimated (Williams and Orville 1988).

To be reliable, all of these static-field measuring instruments must be well maintained since leakage paths to ground, due to dirt and insects, act as a shunt and so weaken or destroy the signal.

Direction-of-arrival measurement

For the detection of lightning over long distances, automatic direction finders (DFs) are the most widely used method at present, measuring the apparent angle of arrival of the sferic waves, several DF stations suitably spaced geographically allowing an intersection point to be computed by triangulation.

Radio waves, like all electromagnetic radiation, are made up of an electric field at right angles to a magnetic field. At radio frequencies below about 10 MHz the ground acts as a good conductor and it is a law of electromagnetic radiation that the lines of electric field touching a conductor do so at right angles. Since the 'transmitting antenna' (the lightning strike) is usually near-vertical, the sferic is radiated with its electric field vertical (vertically polarised). This is important because of the receiving antenna used (see below).

The range over which lightning can be detected depends on the wavelength used. Very low frequencies (VLFs) (3 to 30 kHz, 100 to 10 km) travel great distances, up to megametres, propagating by way of the ground wave which diffracts over the curved surface of the earth. As the distance increases, however, there is an increased loss of the higher frequencies, causing the waveform to become distorted. At ELFs (3 to 30 Hz, 100 000 to 10 000 km), the range can be extended yet further, but the longer the path the greater the distortion (Heydt and Takeuti 1977, Grandt and Volland 1988). In addition, a sky wave is also propagated and this travels by successive reflections between the ground and the ionosphere. Due to the longer path that the sky wave travels, however, it arrives later than the ground wave, distorting the shape of the composite received pulse. Additionally the sky wave is not a single wave following one path, but many waves, due to reflections taking place over a range of angles and through the depth of the ionosphere, which is made up of several layers. In addition, the layers also vary during the day and night. The composite received wave is thus distorted, in part due to its being a mix of ground and sky waves and in part due to the loss of the higher frequencies.

I mentioned earlier that the direction of polarisation of the wave was significant. This is because lightning DF systems use the directional properties of a *loop antenna*, which senses the magnetic (horizontal) component of the sferic wave. A loop aerial is a closed-circuit antenna in the form of a single turn, or a coil, of wire on a frame, often square in shape. Such antennas are directional in sensitivity, with a maximum output when the incoming wave is in line with the frame, falling away to near zero at right angles to the coil. If two antennas are set at right angles to each other, oriented north–south and east–west, the ratio of their signals is proportional to the tangent of the angle of arrival of the sferic.

Of course there is an ambiguity in this arrangement because the signal could be coming from one of two directions, 180 degrees apart. To overcome this, a flat plate antenna is added that detects the vertical direction of the electric-field component and thus the polarity of the discharge. The electric-field component of a wave can point either upwards or downwards, depending on the direction of the current in the transmitting antenna. This information automatically gives an indication of whether the magnetic component is left- or right-handed, and this removes the 180° uncertainty.

The UK Met Office operated the first such direction-finding system from the 1940s to the late 1980s, known as the Cathode Ray Direction Finder (CRDF), working with a centre-frequency of 9 kHz and a bandwidth of 250 Hz. This frequency was chosen because it is roughly the band in which the sferic spectrum is of maximum strength (Ockenden 1947, Maidens 1953, Horner 1954). The signals from the two loop antennas were applied to the x and y plates of a cathode-ray tube, a line on the screen indicating the apparent direction of the flash. An operator at the station would measure the direction on the CRT and reported it to Beaufort Park in the UK, where a fix was obtained by triangulation (Keen 1938). These measurements were then disseminated on the World Meteorological Organization's Global Telecommunication System (WMO 2004). Although of limited accuracy, CRDF was able to cover most of Europe, up to a range of 3000 km.

Modern DF systems are the same in principle but improved and automated by extending the bandwidth covered, through better communication from outstations to base, due to the arrival of computers, and through an increased understanding of exactly which point of the sferic signal to use.

A network of DFs needs a minimum of three receiving stations, the location of the flashes being the point at which the bearings of all the stations intersect. The closer the intersection is to the perpendicular the more reliable the location.

It is necessary to be certain that the received signals are indeed occurring at the same instant; otherwise they may be signals from different flashes. If two or more stations receive pulses within a time window of 5–20 ms they are assumed to come from the same sferic. However, DFs can fail to detect the first return stroke, and may report the second flash instead, causing coincidence to be lost. If the window is widened to be 50 ms more flashes are successfully located, but by increasing the time beyond 100 ms false coincidences start to occur. If the telemetry link is fixed ('dedicated'), the coincidence of flashes is determined from the time of arrival of the data from the DFs, since the time delay in the communication link is constant and known. If the telemetry link introduces any delay, however, such as can occur in a packet-switched communications link, the time of arrival of the sferic at the DFs as indicated by the DF clocks must be used, not the time it arrives at the base station. When two or more stations indicate a coincident event, its position is calculated and its time recorded along with the amplitude and polarity of the peak of the waveform and the number of strikes per flash. The larger the distances covered, the more flashes received, and thus the greater the possibility of confusion over coincidence.

With the minimum configuration of three stations, they are best placed in the corners of an equilateral triangle to reduce the possibility of near-parallel bearings. If four stations are installed, a square is best. For a larger network, stations lying on the same straight line should be avoided. The spacing of the stations should be even, and to take best advantage of the nominal 400 km range of the stations, distances between adjacent sites should be around 150–250 km. The number of stations used, compared with the minimum essential (the redundancy of the network), can be usefully increased, to guard against the failure of a station or of its communications link. Additional stations also give a more accurate fix.

Since the flashes are located by measuring the direction that the sferic signals appear to arrive from, any distortion in this angle leads to an error in flash location. Determining the site errors is, therefore, an important calibration procedure following installation. The distortions vary with direction and in a systematic way (MacGorman and Rust 1988). Once established, corrections can be introduced into the processing software as constants. Before correction, errors as high as ten degrees may occur in some directions.

Random errors can also arise. These may be due to variations from the vertical of the flash and from pulse-distortion during propagation, for example due to mountains. Some of these difficulties were overcome (Krider *et al.* 1976) by gating the sferic analyser to measure the ratio of the two loop-antenna signals during the first 1–5 microseconds of the return stroke, occurring when the strike is about 100 m from the ground and usually nearly vertical, minimising the polarisation error. The short 5-microsecond window also helps to exclude the delayed sky waves, which arrive slightly later, and which cause distortion. Because the high frequencies of the ground wave are attenuated increasingly as they propagate over the curved surface of the earth, the range of accurate detection has to be limited to about 400 km.

The accuracy of strike location varies depending on many factors, but in general, close to the network, strikes can be located to 0–10 km, while at ranges up to 250 km accuracy reduces to 20–30 km because of the uncertainty of the exact angle of arrival. Bearing error is around 1–2 degrees, which gives an increasing error the greater the distance.

Another important accuracy characteristic of a lightning detection network is how effectively it detects the flashes. In tests carried out by MacGorman and Rust (1988), the detection efficiency of the DF systems (with a 100 ms window) was 60–70% at short range and also at longer range in one direction, but only 40–45% at 250 km range in another direction. There was also considerable variability from one time to another, depending, it seems, in part, on the type of lightning and the amount of intra-cloud flashes. (Mach *et al.* 1986). Detection efficiency thus varies from 40 to 70%.

In these tests, ground-truth measurements were obtained by operating video cameras at several locations, the downward-looking camera seeing an all-azimuth view reflected from a conical mirror beneath it. Time was synchronised to 1 ms from an accurate standard and encoded in the video images.

Time-of-arrival measurement

An alternative to measuring the direction of arrival of the sferic signals is to measure the difference in their time of arrival (TOA) at several stations, The position of the flash is then identified by a technique similar to hyperbolic navigation (Lee 1986) by solving for the intersection of the corresponding hyperbolas on a spherical surface. Details of how this process works are given below in the section on arrival-time difference and in Fig. 6.2. An advantage of this method is that whereas the apparent direction of arrival of a sferic wave can be distorted by geographical and man-made features, the time of arrival is less in error, making TOA potentially more accurate. First developed in the early 1980s, two such systems are in operation at present, one being a short-range (400 km) regional system and the other a long-range (12 000+ km) network developed by the UK Met Office (Lee 1989).

A short-range system

A vertical omni-directional whip antenna is all that is required for this type of receiver, sensing the (vertical) electric field of the sferic by capacitive coupling, in contrast to the twin loop antennas measuring the magnetic component by induction, as used in the DF systems. The stations include a receiver and a time generator, which needs to be accurate to a few tenths of a microsecond; this is achieved today by using the Global Positioning System (GPS) signals.

Operating in the 2–500 kHz range, the lightning strike is detected at the field station by similar electronics to that in a DF system, the arrival time of the initial peak of the waveform being determined by the clock. As with the DF receiver, gating the system to look at just the start of the pulse (the first 5 microseconds)

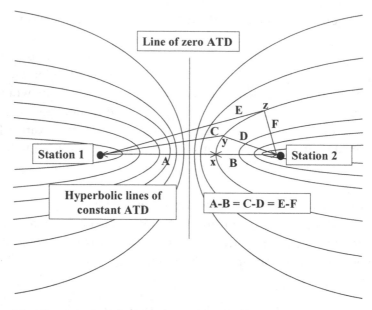

Figure 6.2 The arrival-time-difference (ATD) method of measuring the location of a lightning flash needs a minimum of four receivers. To illustrate the principle, two such stations are shown in the figure. The difference in the time of arrival of a received sferic at the two stations places it somewhere on a parabolic curve (x, y, z, etc). A third receiver would place it precisely on that curve, but there would be two choices of location. The fourth station overcomes this ambiguity. The UK Met Office operates five stations for increased precision and in case of the failure of a station.

prevents sky-wave distortion. But over distances greater than about 400 km, the waveform degenerates too much for accurate timing because of the attenuation of the higher frequencies.

The geometry of the networks is the same as for DF systems, with similar separations of 150–250 km between stations. However, because of the way in which the flash location is derived from the time signals, at least four stations are required (see below). Because the time of arrival of the pulse is not affected by local structures, the local site is less critical.

Errors are introduced, however, by anything that affects the timing of the peak of the signal. Random errors are caused by ambiguity in deciding on the exact position of the peak of the waveform or by drifts in the clock circuits, which can amount to 1 microsecond. Systematic errors are introduced by propagation effects, on both the incoming timing signals (GPS) and the flash waveform, the latter being affected by an increase in rise time caused by the loss of the higher frequencies and variations in the conductivity of the ground (farmland, desert or ocean) over which the wave passes, which affect the speed of the wave slightly. Intervening mountains can also affect path length through propagation delays, amounting to as much as 1.5 microseconds. If these various conditions are different between the flash and the several receivers, different delays will be introduced at each station, with a corresponding error in estimating the difference in arrival times (MacGorman and Rust 1988).

Overall, the detection efficiency of TOA systems varies between 40 and 60%, but there are many caveats to this and it can vary more widely still. There are too many variables to give a single precise figure; in one evaluation of a six-station network, reports suggest that the detection efficiency is 80–85%. Location accuracy is, likewise, widely variable. However, tests seem to have shown that accuracy is the reverse of the DF system, in that location errors peak at 10–20 km when flashes are near to the network, while they fall to 0–10 km when the flash is 250 km from the nearest field station.

Arrival-time difference: a long-range system

The Arrival Time Difference (ATD) network was developed by the UK Met Office as a replacement for the ageing CRDF system, to cover all of Europe and the eastern Atlantic. It has now been in operation since 1988 (Lee 1986, Taylor 1996). TOA systems (as above) did not have sufficiently long range to cover the area required without the use of a large number of stations and so high cost. Radar and satellite measurements were not suitable as an alternative (Lyons *et al.* 1985) (see also below).

The ATD hardware is similar to that of the TOA system and again uses the same ELF/VLF band, so as to obtain long-range measurements. However, to avoid the problems of pulse distortion, inherent in long-range propagation beyond 400 km, the whole waveform is used instead of limiting attention to the first 5 microseconds of the pulse, thereby avoiding having to identify individual features within the pulse waveform. (The difference in names – TOA versus ATD – is not significant: they are only used to allow the two methods to be differentiated from each other.)

The receiver output is sampled every 10 microseconds, and digitised, the most recent 1024 samples being stored. When a signal judged to be sferic-like is received, sampling continues for a further 512 samples and then stops, freezing the waveform. The time difference between the same sferic signals received at different outstations is then estimated by aligning the waveforms in a 'best fit' sense (Lee 1986).

However, the difference in arrival time at two stations does not indicate how far away the flash is. For example if both stations were the same distance from the flash, the ATD would be zero, however close or distant the flash was. But ATD does define a hyperbolic curve along which the flash could occur in order to give the observed ATD (Fig. 6.2). A third station gives a second curve of equal ATD relative to one of the first two stations. Where they intersect indicates the flash location. However, the curves intersect at two points and a fourth station, and its hyperbolic curve, is needed to select which of the two intersections is 'correct'.

The Met Office network has five receivers in the UK, spaced between 300 and 900 km, with one in Gibraltar and one in Cyprus, 1700 and 3000 km distant respectively. Because of the ability of the system to operate over long distances, fewer stations are also required, minimising costs.

Evaluation of location accuracy and detection efficiency depends on having accurate ground-truth data. This is not possible to achieve across the large area covered, but it is feasible to monitor selected areas carefully, for example by using the video system described earlier, and to use this information to assess performance. It is not, therefore, possible to evaluate performance precisely, and so limited case studies must suffice, such as those carried out in France (Massif Central) for the early tests on the ATD system (Lee 1990). From these tests, the location accuracy is currently estimated to be 1–2 km in the UK, 2–5 km in Europe and 5–10 km over the Atlantic. More distant flashes (as far as 12 000 km) can be located to 1–2% of the range. The detection efficiency is around 70%.

A VHF system

A system operating in the much higher VHF band was developed by the French National Aerospace Research Agency in the 1980s. Known as SAFIR, the system uses an array of five vertically polarised dipole antennas spaced around a mast which act as an interferometer to compute the direction and altitude of the flash by sensing the phase difference of the wave arriving at each dipole (Richard and Auffray 1985, Richard *et al.* 1988). It operates in the 110–118 MHz band, and so is limited to line-of-sight distances. However, the measurement is very precise since the waveform is not modified by ground attenuation or by interfering sky waves and also because it is the phase, not the shape of the waves, that is used. Manufacturers claim better than 1 km location accuracy and over 90% detection efficiency. As with VLF DF systems, the VHF stations are deployed on a triangular grid, in this case 20–100 km distant from a central processing station, giving a range of location of about 200 km. At these frequencies it is possible to detect the initial cloud-to-ground stepped leader as well as the main return strikes, in addition to lightning within clouds, the latter often being a precursor of ground strikes (Richard *et al.* 1988).

The system also includes a conventional VLF direction finder working in the range 300 Hz to 3 MHz. Proctor (1971) and Krehbiel *et al.* (1984) describe a VHF system using the ATD method, although its range was limited because of the line-of-sight nature of VHF waves.

Radar measurements

As will be discussed in Chapter 10, radar can be used to measure rainfall. By applying the same techniques to examine thunderclouds, the evolution of the cloud can be followed closely. If two Doppler radars are used, it is possible to measure the changing winds within the cloud by sensing the velocity of the precipitation particles towards or away from the radars. The returned signal can locate areas of high precipitation, while measurement of the angle of polarisation gives information on the shape and fall-mode of the particles, and thereby information on the types of particle. From such measurements it has been found that hail is often present in those parts of the cloud where lightning is initiated.

If the radar pulse is circularly polarised, the returned signals indicate the hand-edness of the returned waves' polarisation and this tells us something about the alignment of the particles. In a thundercloud the main determiner of particle align-ment is the electric field, and so it is possible to watch the build-up of electric charge and its collapse after the lightning strike.

Although radar can be used as a lightning warning system, it is most useful for investigating the processes of thundercloud electrification (Houze 1993, Rutledge and Petersen 1994, Samsury and Orville 1994, Zipser and Lutz 1994, Mohr and Torancinta 1996). It also has advantages over the use of research aircraft for this purpose, being able to obtain a three-dimensional image with a resolution of about 100 m every few minutes.

Satellite measurements

Although systems such as the ATD network can measure sferic activity at great range, there are large areas which cannot be covered by the shorter-range methods. In remote areas such as these, satellite observations play an important role. The sensing of lightning from space is new, a Lightning Imaging Sensor (LIS) having been launched in November 1997 aboard the Tropical Rainfall Measuring Mission (TRMM) satellite which is in a nearly circular orbit inclined at 35 degrees at an altitude of 350 km (see Chapter 11 for details of the TRMM satellite). There is a similar device known as an Optical Transient Detector (OTD). The LIS comprises what is termed a *staring imager*, a camera with a wide field of view, a narrow-band filter centred at 777 nm (near infrared) with a high-speed charge-coupled-device (CCD) sensor. A processor determines in real time when a lightning flash has occurred. It can detect this easily at night, the flash spreading up through the cloud and laterally.

In daylight pulses are more difficult to detect, but the processor goes some way to remove the background signal, allowing about 90% detection rate. Only rapid changes of illumination have to be detected, so the constant background can be removed. It is able to resolve storms 3–6 km in scale over an area of 550 × 550 km at the earth's surface. The satellite is, of course, moving relative to the ground at a speed that allows any one point to be in the field of view of the camera for just 80 seconds. Any flashes occurring during this time are detected, the radiant energy being measured and the location estimated (Christian *et al.* 1992). The records from the LIS are compared with ground-truth measurements (made with ground-based lightning detectors). Comparison can also be made with cloud images and rain measurements made both on the ground and from satellites. In this way the correlation of lightning with other phenomena can be studied while also checking the efficiency and accuracy of the LIS.

As the satellite orbits the globe, a map is built up of lightning flashes. Of course a snapshot impression of just over a minute is not sufficient to give very extensive coverage and it will miss many storms that come and go while the camera is not overhead, but it is a start, and gives a summary of activity if not full cover. Indeed, bearing in mind the number of strikes occurring worldwide at any one time, full cover might prove difficult. A system for use on geosynchronous satellites is being developed (it is understood) which would give continual cover (Christian *et al.* 1989). At this distance the signals will be that much smaller and more difficult to detect and there will be many more signals to detect.

A number of false strikes can be indicated since the LIS also responds to non-lightning and even non-optical signals. Software filtering has been developed to remove many of these, which can be caused by solar glint off the sea or lakes, electronic noise in the system (that has to detect very small light pulses), or even energetic particles in the Van Allen radiation belts striking the CCD sensing surface. As with so much work in satellite sensor development, the pace of change is high and the instruments very specialised. For those interested, the best up-to-date information is to be found on NASA websites.

References

Christian, H. J., Blakeslee, S. J. and Goodman, S. J. (1989). The detection of lightning from geostationary orbit. *Journal of Geophysical Research*, **94**, 13329–13337.

Christian, H. J., Blakeslee, S. J. and Goodman, S. J. (1992). *Lightning Imaging Sensor (LIS) for the Earth Observing System.* NASA Tech. Mem. 4350, MSFC, Huntsville, AL.

Dudhia, J. (1996). Back to basics: thunderstorms, part 1. *Weather*, **51**, 371–376.

Dudhia, J. (1997). Back to basics: thunderstorms, part 2. Storm types and associated weather. *Weather*, **52**, 2–7.

Durand, M. and Wilson, J. G. (2006). Ball lightning and fireballs during volcanic air pollution. *Weather* **61**, 40–43.

Gosline, A. (2005). Lightning: thunderbolts from space. *New Scientist,* No. 2498, 30–34.

Grandt, C. and Volland, H. (1988). Locating thunderstorms in South Africa with VLF sferics: comparison with METEOSAT infrared data. In *Proceedings of the Eighth International Conference on Atmospheric Electricity.* Uppsala, Sweden: Institute of High Voltage Research, pp. 660–666.

Hardaker, P. J. (1998). Lightning never strikes in the same place twice – or does it? In *Lightning Protection 98: Buildings, structures and electronic equipment.* Birmingham: ERA Technology.

Heydt, G. and Takeuti, T. (1977). Results of the global VLF-atmospherics analyser network. In *Electrical Processes in Atmospheres* (ed. H. Dolezalek and R. Reiter). Darmstadt: Steinkopf, pp. 687–692.

Horner, F. (1954). New design of radio direction finder for locating thunderstorms. *Meteorological Magazine,* **83**, 137–138.

Houze R. A., Jr. (1993). *Cloud Dynamics.* San Diego, CA: Academic Press.

Johnson, R. L. and Janota, D. E. (1982). An operational comparison of lightning warning systems. *Journal of Applied Meteorology,* **21**, 703–707.

Keen, R. (1938). *Wireless Direction Finding.* London: Iliffe and Sons; Andover: Chapel River Press.

Krehbiel, P. R., Brook, M., Khanna-Gupta, S., Lennon, C. L. and Lhermitte, R. (1984). Some results concerning VHF lightning radiation from the real-time LDAR system at KSC, Florida. In *Seventh International Conference on Atmospheric Electricity.* Boston, MA: American Meteorological Society, pp. 388–393.

Krider, E. P., Noggle, R. C. and Uman, M. A. (1976). A gated wideband magnetic direction finder for lightning return strokes. *Journal of Applied Meteorology,* **15**, 301–306.

Lee, A. C. L. (1986). An experimental study of the remote location of lightning flashes using a (VLF) arrival time difference technique. *Quarterly Journal of the Royal Meteorological Society,* **112**, 203–229.

Lee, A. C. L. (1989). Ground truth confirmation and theoretical limits of an experimental VLF arrival time difference lightning flash location system. *Quarterly Journal of the Royal Meteorological Society,* **115**, 1147–1166.

Lee, A. C. L. (1990). Bias elimination and scatter in lightning location by the (VLF) arrival time difference technique. *Journal of Atmospheric and Oceanic Technology,* **7**, 719–733.

Lyons, W. A., Bent, R. B. and Highlands, W. H. (1985). Operational uses of data from several lightning position and tracking systems (LPATS). In *Tenth International Aerospace and Ground Conference on Lightning and Static Electricity, Paris, June 10–13,* Paris: AAAF. pp. 347–356.

MacGorman, D. R. and Rust, W. D. (1988). An evaluation of the LLP and LPATS lightning ground strike mapping systems. In *Proceedings of the Eighth International Conference on Atmospheric Electricity.* Uppsala, Sweden: Institute of High Voltage Research, pp. 668–673.

Mach, D. M., MacGorman, D. R., Rust, W. D. and Arnold, R. T. (1986). Site errors and detection efficiency in a magnetic direction-finder network for locating lightning strokes to ground. *Journal of Atmospheric and Oceanic Technology,* **3**, 67–74.

Maidens, A. L. (1953). Methods of synchronising the observations of a sferics network. *Meteorological Magazine,* **82**, 267–270.

Mohr, K. I. and Torancinta, E. R. (1996). A comparison of WSR-88D reflectivity, SSM/I brightness temperature and lightning for Mesoscale Convection System in Texas.

Part II: SSM/I brightness temperature and lightning. *Journal of Applied Meteorology*, **35**, 919–931.

Muir, H. (2001). Puzzle balls. *New Scientist*, No. 2322/23, 12.

Ockenden, C. V. (1947). Sferics. *Meteorological Magazine*, **76**, 78–84.

Proctor, D. E. (1971). A hyperbolic system for obtaining VHF radio pictures of lightning. *Journal of Geophysical Research*, **76**, 1478–1489.

Richard, P. and Auffray, G. (1985). VHF–UHF interferometric measurements: applications to lightning discharge mapping. *Radio Science*, **20**, 171–192.

Richard, P., Soulage, P., Laroche, P. and Appel, J. (1988). The SAFIR lightning monitoring and warning system: application to aerospace activities. In *International Aerospace and Ground Conference on Lightning and Static Electricity*. Oklahoma City, OK: NOAA. pp. 383–390.

Royal Meteorological Society (1994). Scientific statement on lightning. *Weather*, **49**, 27–33.

Royal Society (2002). Ball lightning. Compiled by J. Abrahamson. *Philosophical Transaction of the Royal Society A*, **360** (1790).

Rutledge, S. A. and Petersen, W. A. (1994). Vertical radar reflectivity structure and cloud-to-ground lightning in the stratiform region of Mesoscale Convective Systems: further evidence for in situ charging in the stratiform region. *Monthly Weather Review*, **122**, 1760–1766.

Samsury, C. E. and Orville, R. E. (1994). Cloud-to-ground lightning in tropical cyclones: a study of hurricanes Hugo (1989) and Jerry (1989). *Monthly Weather Review*, **122**, 1887–1896.

Taylor, P. (1996). When lightning strikes: split-second timing and thunderstorm forecasting. *GPS World*, November 1996, 24–32.

Williams, E. R. and Orville, R. E. (1988). Intracloud lightning as a precursor to thunderstorm microbursts. In *International Aerospace and Ground Conference on Lightning and Static Electricity*. Oklahoma City, OK: NOAA. pp. 454–259.

Williams, H. (2001). Rider on the storm. *New Scientist*, No. 2321, 36–40.

WMO (1996). *Guide to Meteorological Instruments and Methods of Observation*, 6th edn. WMO 8. Geneva: WMO.

WMO (2004). *Manual on the Global Telecommunication System*. WMO publication 386. Geneva: WMO.

Zipser, E. J. and Lutz, K. R. (1994). The vertical profile of radar reflectivity of convective cells: a strong indicator of storm intensity and lightning probability? *Monthly Weather Review*, **122**, 1751–1759.

Part 3

Measuring precipitation

Having arrived at an understanding of how precipitation forms, we can move on to see how it can be measured and the many problems that arise in the attempt to obtain precise data. The importance of measurements was stressed by Leonardo da Vinci when he wrote:

No human enquiry is worthy of the name of science unless it comes through mathematical proofs. And if you say that the sciences which begin and end in the mind possess truth, this is not to be conceded, but denied for many reasons. First because in such mental discourses there enters no experiment, without which nothing by itself reaches certitude.

In the above, 'experiment' also had the meaning of 'experience' in Leonardo's time. What he was saying was what others have said often since, that Aristotle's way of doing 'science', by just thinking and talking without looking and testing was not science at all.

Until the late nineteenth century, attempts to measure rain were quite rare, and so each is remarkable and worthy of comment. The next chapter traces the history of these early attempts. Their origins were probably in simple agricultural aids, but in Europe from the seventeenth century onwards there developed a more academic approach, arising from scientific curiosity aided by improving technology. In Chapters 8 to 11, I review the current state of the technology, looking in some detail at how precipitation is measured today with raingauges, radar and satellites. The pace of development is considerable in satellite instrumentation and radar, but even in the old technology of raingauges there remain some difficult problems to be solved. The measurement of snowfall presents particular problems, and has a chapter to itself.

7

Early attempts to measure rainfall

India

The first written reference to the measurement of rainfall appears in India in the fourth century BC, in Kautilya's *Arthasastra* (Shamasastry 1915). From it we read that 'In front of the storehouse, a bowl with its mouth as wide as an Aratni (18 inches) shall be set up as a raingauge (Varshanana).' Later it is said that

The quantity of rain that falls in the country of Jangala is 16 dronas; half as much more in moist countries; $13\frac{1}{2}$ dronas in the country of Asmakas . . . The forecast of such rainfall can be made by observing the position, motion and pregnancy of Jupiter, the rise and set motion of the Venus and the natural or unnatural aspect of the sun . . . Hence, according as the rainfall is more or less, the superintendent shall sow the seeds which require either more or less water.

Palestine

The second known written record of the measurement of precipitation is in the *Mishnah*, a book recording 400 years of Jewish life in Palestine from the early second century BC to the end of the second century AD (Danby 1933). Rainfall is reported for a full year, although it is not clear if this is just one year or the average of many (Vogelstein 1894). The year is divided into three periods. The 'early period', or autumn rain, is from October to mid December, 'which moistens the land and fits it for the reception of the seed'. The 'second period' is from mid December to mid March, during which 'the copious winter rain, which saturates the earth, fills the cisterns and pools and replenishes the springs'. The 'third period', or spring rain, is from mid March to April or May, when the rainy season ends, 'which causes the ears of corn to enlarge, enables the wheat or barley to support the dry heat of the early summer, and without which the harvest fails'.

The rainfall amounts recorded were:

early rain period	1 *tefah*	90 mm
second period	2 *tefah*	180 mm
third period	3 *tefah*	270 mm
annual rainfall	6 *tefah*	540 mm

The book also contains notes on soil moisture, commenting that rainfall percolated to a depth of 1 *tefah* in barren soils, 2 *tefah* in medium soils and 3 *tefah* in broken-up arable lands. Like the Indian measurements, these early rainfall observations in Palestine were intended purely for agricultural use. Present-day annual rainfall in the region varies greatly, being highest in the west at 700 mm and least in the Jordan valley, where on average it is around 150 mm. Not knowing exactly where the ancient measurements were made, a precise comparison is not possible, although the amounts recorded seem of a similar order to today's, and rainfall still follows the same yearly cycle, starting in October and continuing until April.

But neither this nor the Indian attempts continued for long; they were just isolated events, quickly forgotten. And as we saw in Chapter 2, it was to be over a thousand years before there was any further rational or secular enquiry into the workings of the natural world – and this included quantitative hydrological or meteorological measurements.

China

It was in China around AD 1247 that the next known quantitative rainfall measurements were made, described in the mathematical treatise *Shushu jiuzhang* by Qin Jiushao (Ch'in Chiu-Shao). The raingauges were conical or barrel-shaped, one being installed at each provincial and district capital. In addition, the book also discusses problems with large snow gauges made from bamboo which were sited in mountain passes and uplands, probably the first ever reference to snow measurement. Qin Jiushao also discusses how point measurements of rainfall were converted to areal averages (Needham 1959). Biswas (1970) suggests that the gauges were necessary because the flooding of rivers and canals has always been a problem in China, but if this was the case the hydrological cycle must have been understood and this seems improbable in the thirteenth century; more likely the purpose was again agricultural.

Korea

The need for regular rainfall for the cultivation of rice was probably the incentive for the introduction of raingauges into Korea, most likely from China, during the fifteenth century. The economy of the country depended on the crop, and prayers were said for rain to the spirits of the mountains and rivers. The gauges were introduced at the height of Korean civilisation during the reign of King Sejong of the Lee dynasty in 1441 and were used continuously, unchanged, until 1907 (Fig. 7.1).

One of the earliest references to the gauges (Wada 1910) says:

In the 24th year of the reign of King Sejong, the King caused an instrument of bronze to be constructed for measurement of rainfall. It was an urn 1 *shaku*, 5 *sun* deep and 7 *sun* broad (about 30 cm deep and 15 cm in diameter), set on a pillar. The instrument was placed at the observatory and the officials of the observatory measured the depth of rainfall each time it rained. The results were made known to the King. Similar instruments were distributed to the provinces and cantons, and the results of the observations were reported to court.

In Korean history books, the gauges make another appearance two centuries later:

In the year 46 (1770) of the King Eijo, the King, following the ancient system of King Sejong, had numerous raingauges constructed, and placed them, two at the palace, by the side of the wind-vane, and the others on the chief places of the eight provinces. The rain gauges are placed on stone pillars, measuring 1 in height and 0.8 at the side, and on the upper surface is a hole 1 deep to receive the end of the instrument.

This account appears in both Wada (1910) and Anonymous (1911) (note that the dimensions, with no units given, are as they appear in the original). Wada was Director of the Korean Meteorological Observatory in the early twentieth century and searched for early raingauges, finding one in Seoul and one in Taiku, the second one being still in use. There does not appear to be any record of the actual data collected by the gauges.

The reinvention of the raingauge and of the quantitative measurement of rain in China and Korea are probably the greatest achievements in meteorology and hydrology during the 1300 years since the isolated Palestinian and Indian measurements. It is notable that all of these originated in the East, Europe being dominated by religious repression, with people burnt at the stake for daring to question anything.

But during the sixteenth and seventeenth centuries the tables turned, and with the Renaissance and Enlightenment in Europe there was an explosion of inventiveness in ways of measuring the world, paralleling the growing theoretical insights examined in Chapter 2.

As Johannes Kepler pointed out in the early seventeenth century, in the same spirit as Leonardo a hundred years earlier, 'To measure is to know'. This idea is now absolutely basic to science, yet it took so long to come about.

Figure 7.1 Reproduction of a raingauge used widely in Korea from the fifteenth up to the twentieth century. It was probably introduced from China. This replica is at the Science Museum in London and was kindly made available for photographing by Jane Insley.

Europe

Benedetto Castelli (1578–1643)

There were no (known) raingauges in Europe until the seventeenth century, the first recorded instrument being made in Italy by Benedetto Castelli, a Benedictine monk and student of Galileo. In a letter to Galileo in 1639 Castelli says:

Being returned to Perugia, there followed a rain, not very great but constant and even, which lasteth for a space of eight hours or thereabouts; and it came into my thoughts to examine, being in Perugia, how much the lake (Thrasimeno) was increased and raised by this Rain, supposing (as it was probable enough) that the rain had been universal over all the lake; and like to that which fell, in Perugia, and to this purpose I took a glasse formed like a cylinder, about a palme high, and half a palme broad [12 cm diameter]; and having put in it water sufficient to cover the bottom of the glasse, I noted diligently the mark of the height of the water in the glasse, and afterwards exposed to open weather, to receive the rain water, which fell into it; and I let it stand for the space of an hour; and having observed that in that time the water was risen in the vessel the height of the following line (about 10 mm long to represent the depth). I considered that if I had exposed the same rain such other vessel equal to that, the water would have risen in them according to that measure.

This is a correct copy from Castelli (1661), although something seems to have been lost in translation. The meaning is clear enough, however.

But Castelli appears to have made only the one measurement of an isolated event, and seems not to have considered keeping a record over time.

Sir Christopher Wren (1632–1723)

In the 1660s Sir Christopher Wren made the first known British gauge. All we have on its design is a sketch made in 1663 by a visiting French diplomat, Balthasar de Monconys, although it was probably only a rough prototype (Middleton 1968). From the sketch it appears that it was part of a recording 'meteorograph', but the details are too vague to warrant further comment. The sketch can be seen in Biswas (1970).

A Wren–Hooke tipping-bucket raingauge was next developed as part of a 'weather-wiser' described by Grew (1681), a former secretary of the Royal Society. It was driven by a strong, large, weight-driven pendulum clock and recorded barometric pressure, temperature, rain, relative humidity and wind direction. All of these measurements were recorded on paper tape as punched holes every 15 minutes. This is in fact an Automatic Weather Station recording on punch tape, made 350 years ago!

The method of measuring rainfall was also similar to most of today's raingauges, namely a tipping bucket. However, it differs from today's designs in that it had only

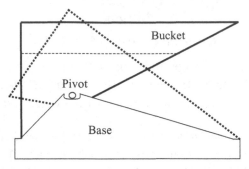

Figure 7.2 Around 1680 Sir Christopher Wren and Robert Hooke designed a tipping bucket to measure rainfall. This figure is my adaptation of a drawing by Hooke (Biswas 1970). The bucket is shown in bold in its level, collecting, position and dotted in its tipped, emptying, position. See text for details of its operation.

one side, rather than two. Robert Hooke made a sketch of the mechanism and Figure 7.2 is my simplified version of the drawing. The bucket required some form of counterbalance (provided in today's designs by having a symmetrical two-sided bucket (see next chapter). This was not illustrated by Hooke but he describes it as being a cylinder immersed in water (or other liquid), the cylinder slowly being raised out of the liquid as the bucket filled, so arranged that when the bucket was filled to its 'designed fullness' it immediately emptied itself and the counterbalance cylinder 'plunged itself into the water and raised the bucket to a place where it was again to begin its descent'.

But the tipping-bucket idea was not new even then, being known to the Arabs and described in a manuscript *al-Jazari* by Muhammad ibn Ibrahim in a thesis on machines around 1364. He describes a mechanism that dispenses wine to two people, alternately, through a double-sided tipping bucket. The idea was also known to the Japanese, who used a bamboo tube that slowly filled with water, the tube tipping when full and making a noise to scare birds off crops. These can be bought in some garden centres today.

Robert Hooke (1635–1703)

Hooke is a much underrated scientist, in large part because Newton disliked him and wrote him out of history after his death. It was probably Hooke rather than Newton, for example, who thought up the idea of the reflecting telescope, although it was Newton who made one and so got the credit. Hooke worked with Wren on the tipping-bucket gauge and also made and used a non-recording raingauge consisting of a large glass bottle holding 2 gallons (9 litres) of water held in a wooden frame with a glass funnel of 11.4 inches (29 cm) held in place against the wind by stays.

The neck of the container was long and of small bore so as to minimise evaporation. It was operated throughout 1695 at Gresham College, where Hooke was professor of geometry, being read every Monday. Over the year it collected 29 inches (737 mm) of rain. But none of these gauges operated for extended periods.

Richard Townley (1629–1707)

The first continuous record of rainfall was made by Richard Townley, in Lancashire, from 1677 to 1703 (Townley 1694). The gauge was a 12-inch (30 cm) diameter funnel on the roof of his house connected by an 8-m lead pipe to a graduated cylinder inside.

Interest in measuring rainfall increased rapidly in the eighteenth century, worldwide, and so only the key points can be recorded from this point on.

Dr D. Dobson

I cannot find Dobson's first name or dates of birth and death, but he was mentioned earlier, in Chapter 2, with regard to tests of evaporation in a vacuum. These experiments may throw some doubt on either his competence or his honesty, but nevertheless he was among the first to expose a raingauge to today's standards on a large, open, grassy patch. Most gauges until this time had been on roofs, it being thought that this would allow them to 'record free fall of rain' (not yet realising the problem of wind effects). Dobson's site (1777) overlooked Liverpool and was well exposed all round, the gauge being 12 inches (30 cm) in diameter and part of a larger experiment concerned with evaporation. So the first modern long-term rainfall measurements, with raingauges exposed as they are today, were made amidst the industrial revolution in the north of England, the seat of the modern industrial world. I grew up near Liverpool amidst the remnants of these heavy industries, with tall factory chimneys, coal mines and steam trains, playing rugby in the fog on cold Wednesdays at my grammar school – quite atmospheric with hindsight, but not at the time.

Gilbert White (1720–1793)

Around the middle of the eighteenth century, with long-term rainfall records now being kept, the question of mean annual rainfall became of interest. Gilbert White is well known for his *Natural History of Selborne* (1789), and his house and garden are still there and open to the public in this quiet Hampshire village. Among the many other observations of the natural world made by White was a series of rainfall readings from May 1779 until December 1786 (Table 7.1), although he considered

Table 7.1 *Rainfall measured by Gilbert White at Selborne.*

Year	Total (inches)
1780	27.32
1781	30.71
1782	50.26
1783	33.71
1784	38.80
1785	31.55
1786	39.57

1 inch = 25.4 mm.

this not long enough to get a good estimate of mean rainfall. The average for this period was 35.99 inches (914 mm), ranging from a minimum of 27.32 (693.9 mm) to a maximum of 50.26 (1276.6 mm).

The longest record by one person, using the same instrument, at this time was kept by White's brother-in-law, Thomas Barker of Lyndon in Rutland, who made observation for 59 years (1736–1796). He commented to White (1789) that 'The mean rain of any place cannot be ascertained till a person has measured it for a very long period', and went on to give some actual figures:

If I had only measured the rain for the four first years from 1740 to 1743, I should have said the mean rain at Lyndon was 16 and a half inches [419.1 mm] for the year; if from 1740 to 1750, 18 and a half inches [469.9]. The mean rain before 1763 was 20 and a quarter [514.4], from 1763 and since 25 and a half [647.7]; from 1770 to 1780, 26 [660.4]. If only 1773, 1774 and 1775 had been measured, Lyndon mean rain would have been called 32 inches [812.8 mm].

There will be more on the need for long runs of data to get accurate means in Chapters 11 and 12.

William Heberden (1710–1801)

As described in Chapter 2, William Heberden (1769) was the first to observe how raingauge exposure affected the gauge's catch when he noticed that two identical gauges in London a mile apart caught different amounts of rain, the one fixed above all surrounding houses always catching less than one lower down. To investigate this, Heberden installed two gauges of his own, one on the chimney of his house, the other in his garden. He also installed a third gauge on the 150-foot (45 m) tower of Westminster Abbey. He took readings every month for a year and they showed that the gauge on the chimney caught only 80% of that in the garden, while the gauge on the tower caught only just over 50% (Reynolds 1965).

He could not explain why this was, but speculated that it might be some electrical effect. A view at the time was that raindrops might increase in size by condensation as they neared the ground. It will be recalled that the processes by which raindrops form and the nature of cloud droplets were still not known for certain at this time (see Chapter 2). But doubts arose about the idea that the drops grew as they fell the last few hundred metres and then, in 1822, Henry Boase of Penzance noticed that, for some reason, the difference in catch between the higher and lower gauges increased as the windspeed increased. He speculated that it might be due to wind eddies, and that gauges should be sunk to ground level. No one paid any attention to this and everyone continued to believe that the drops increased in size as they fell the last few hundred metres. Others suggested that it was a temperature effect because the differences were greatest in winter. But the matter was finally dealt with decisively by William Jevons.

William Stanley Jevons (1835–1882)

Jevons (1861) describes experiments in which he constructed a small wind tunnel with glass walls wherein he placed objects. To show the stream lines he used smouldering brown paper to make smoke trails and from these tests drew several figures, one (Fig. 7.3) being the flow of air over a raingauge. He concludes his (very clearly argued) paper by saying 'My conclusions, shortly stated, are:

(1) An increase of the rainfall close to the earth's surface is incompatible with physical facts and laws.
(2) The individual observations on this subject are utterly discordant and devoid of law when separately examined, and the process of taking an average under such circumstances gives an apparent uniformity which is entirely falacious.
(3) When daily measurements of rain, or even monthly totals, are examined with reference to the strength of the wind at the time, it becomes obvious that there is a connection.
(4) Wind must move with increased velocity in passing over an obstacle. It follows demonstratively that rain-drops falling through such wind upon the windward part of the obstacle will be further apart, in horizontal distance, than where the wind is undisturbed and of ordinary velocity.'

George James Symons (1838–1900)

A contemporary of Jevons, George Symons was a young assistant in the newly formed Meteorological Department of the Board of Trade (the first UK Met Office). He became interested in rainfall and its measurement, fired by the drought years

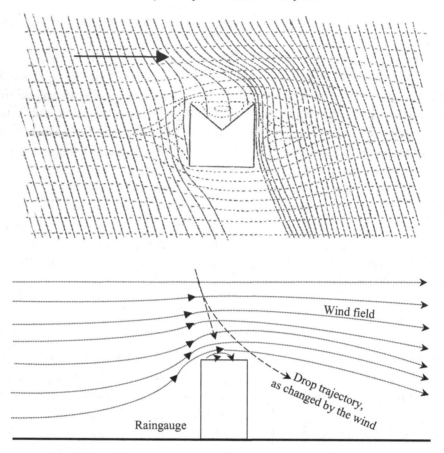

Figure 7.3 William Stanley Jevons was the first person to explore in detail the effects of wind flow over and around a raingauge, to explain the loss of raindrops due to the distortion of the wind field caused by the presence of the gauge. The top drawing is by Jevons from his paper of 1861, the lower illustration being my summary. However, as is shown in the next chapter, the flow of air in the real world is not as simple as either drawing suggests.

of 1854–1858, and it became a lifetime commitment and a personal crusade. He amassed rainfall data from many sources and experimented with the various gauges then in use.

In 1860–1861 he published the first volume of *British Rainfall*, which contained 168 annual totals, mostly from southern England but also from Guernsey and the Isle of Man (Symons 1864, 1866). He resigned his post at the Board of Trade in 1863 to devote all his time to the task of improving rainfall measurement and data collection, much to the benefit of meteorology but at the cost of great financial difficulties for himself.

In the twentieth annual volume of *British Rainfall* Symons wrote, 'There is no point in the study of rainfall of greater interest and practical utility than the accurate determination of the average annual fall'. He also realised that it was very easy to make approximate measurements of rainfall but that it got progressively harder to improve on this, and that it was extremely difficult to obtain and verify an absolute measurement. This opinion led Symons to investigate the wind problem further and in 1881 he confirmed, through many experiments, that wind was indeed the cause of reduced catch with height, and showed, using a gauge 12 inches (30 cm) high as the reference, that a gauge at only 2 inches (5 cm) above the ground caught an extra 5% while gauges at 5 and 20 feet (1.5 and 6 m) lost 5% and 10% respectively (Kurtyka 1953). These were averages and will vary from one rain event to another depending on windspeed and drop size. In 1893 Abbé showed that the effects of wind depend on the texture of the rain, being more pronounced in drizzle than for larger drops. Wind effects are very important and are considered in detail in Chapter 8.

By the time Symons died in 1900, he was receiving records from 3500 sites. Symons was one of the most significant figures there has ever been in rainfall measurement and the collection of rainfall data. The science is much in his debt.

This is a good place to stop and move on to the next chapter, to look at how raingauges developed during the twentieth century.

References

Anonymous (1911). Korean rain gauges of the fifteenth century. *Quarterly Journal of the Royal Meteorological Society*, **37**, 83–86.

Biswas, A. K. (1970). *History of Hydrology*. London: North-Holland.

Castelli, B. (1661). *Of the Mensuration of Running Water*, translated by T. Salisbury. London: William Leybourn, p. 28.

Danby, H. (1933). Translation of *The Mishnah*. Oxford: Oxford University Press.

Dobson, D. (1777). Observations on the annual evaporation at Liverpool in Lancashire; and on evaporation considered as a test of the dryness of the atmosphere. *Philosophical Transactions of the Royal Society of London*, **67**, 244–259.

Grew, N. (1681). In *Musaeum Regalis Societatis*. London: W. Rawlins, pp. 357–358.

Heberden, W. (1769). Of the quantities of rain which appear to fall at different heights over the same spot of ground. *Philosophical Transactions of the Royal Society of London*, **59**, 359.

Jevons, W. S. (1861). On the deficiency of rain in an elevated rain-gauge, as caused by wind. *London, Edinburgh and Dublin Philosophical Magazine*, **22**, 421–433.

Kurtyka, J. C. (1953). *Precipitation Measurement Study*. Report of Investigation 20, State Water Survey, Illinois. Urbana, IL: Department of Registration and Education.

Middleton, W. E. K. (1968). *Invention of the Meteorological Instruments*. Baltimore, MD: Johns Hopkins Press.

Needham, J. (1959). *Science and Civilisation in China, Vol 1*. Cambridge: Cambridge University Press.

Reynolds, G. (1965). A history of raingauges. *Weather*, **20**, 106–114.

Shamasastry, R. (1915). Translation of *Kautilya's Arthasastra*. Government Oriental Library Series, Bibliotheca Sanskrita, No. 37, part 2. Bangalore.

Symons, G. J. (1864). Rain gauges and hints on observing them. *British Rainfall*, 8–13.

Symons, G. J. (1866). On the rainfall of the British Isles. In *Report on 35th Meeting of the British Association for the Advancement of Science*, Birmingham, 1865, pp. 192–242.

Townley, R. (1694). Observations on the quantity of rain falling monthly for several years successively: a letter from Richard Townley. *Philosophical Transactions of the Royal Society of London*, **18**, 52.

Vogelstein, H. (1894). Die Landwirthschaft in Palästina zur Zeit der Mišnah. Part 1, Getreidebau. Dissertation, Breslau.

Wada, Y. (1910). *Scientific Memoirs of the Korean Meteorological Observatory, Vol 1*. Chemulpo.

White, G. (1789). *The Natural History and Antiquities of Selborne*. London. Available from many publishers, such as Penguin Books, Thames & Hudson, etc.

8

Measuring precipitation with raingauges

By the mid nineteenth century all basic raingauge design principles had been established, in no small part thanks to George Symons. We will now take a look at the design of modern raingauges and then at what errors need to be taken note of. Because the measurement of snowfall is even more difficult than the measurement of rainfall, for reasons that will be made clear later, a separate chapter (Chapter 9) is devoted to it.

Definitions

By 'raingauge' I mean a device that collects the rain in a funnel and then measures the water in some way. A few raingauges do not actually catch the rain but measure it indirectly, and these are included at the end of the chapter.

The term 'precipitation' includes rain, drizzle, snow and hail, but not condensation in the form of dew, fog, hoar frost or rime, even though they can produce trace readings in a gauge of up to 0.2 mm. The total precipitation is the sum of all the liquid collected (including the water produced from melted solid precipitation), expressed as the depth it would cover on a flat surface assuming no losses due to evaporation, runoff or percolation into the ground. While inches have been used in the past to measure precipitation, and still are in some countries, millimetres are the more usual unit today (0.01 inches is about equivalent to 0.25 millimetres).

In this chapter we will look firstly at some actual raingauges – manual, mechanical and electronic – and then at their faults and errors and at what can be done to lessen these.

Manual raingauges

I do not want to dwell too long on constructional details but simply to outline the general principles of gauges. Sevruk and Klemm (1989) have estimated that there are over 50 different types of manual gauge in general use worldwide, but as they all have much in common it is only necessary to describe a typical example, and the UK case will be used.

The UK Meteorological Office Mark 2 raingauge

This classic gauge demonstrates the general features of most manual gauges and is based on Symons's 5-inch model (12.70 cm diameter) with an area of 126.7 cm^2 (Fig. 8.1). It has been the UK standard manual gauge for all of the twentieth century and it continues into the twenty-first.

A copper funnel with a precisely turned bevelled brass rim fits onto a copper base set into the ground, the base splayed out to give it more stability. It has deep sides to minimise outsplash. Inside the base is a removable copper container and inside the container is a bottle. The gauge is installed with its rim 12 inches (30 cm) above the ground, which should be of short-cut grass or, failing this, gravel. A hard smooth surface should not be used, to avoid insplash. The bottle holds the equivalent of 75 mm (3 inches) of rain. The removable copper container provides extra capacity in case the bottle overflows in heavy rain or the gauge is left too long, the two together holding 140 mm (5.5 inches). A cheaper model, known as the Snowdon pattern, has straight sides and so is less stable. A modified version of the Snowdon gauge, known as the Bradford gauge, is deeper and can hold 680 mm of rain. A still cheaper version is made in galvanised steel. Another cheaper but larger gauge has a funnel aperture of 20.32 cm (8 inches) and a total measuring capacity of 180 mm of rain (7 inches). In the USA, 8 inches is most commonly used.

The gauges are read daily using a graduated glass cylinder which is either flat-based or tapered, the latter (as illustrated) for greater accuracy when small amounts of water are collected. The cylinders are marked in steps of 0.1 mm with an additional mark at 0.05 mm. Weighing the water gives a more accurate measure if needed.

When gauges cannot be read daily, a larger container is necessary so that weekly or monthly amounts of rain can be stored. In the UK two such models are used, both having a funnel diameter of 12.70 cm, one holding 680 mm of rain (26.8 inches) and a larger model holding 1270 mm (50 inches), the latter for areas of high rainfall or where the gauge may have to be left for up to two months. It is known as the Octapent gauge because it is a merging of a 5-inch funnel with the larger base of an 8-inch gauge.

These two gauges have a precisely fitting removable inner container to hold the water. The amount of water is first estimated in the container by means of a dipstick, then measured more precisely using a graduated cylinder.

Figure 8.1 The UK Met Office standard manual raingauge is the 5-inch Mark 2. The collected water is stored in a bottle within the gauge and is measured daily with the graduated cylinder. The cylinder tapers at the bottom to help measure small amounts more accurately. The base of the gauge extends below the ground and splays out to give stability.

Other types of manual raingauge

There are more than 150 000 manually read gauges in use throughout the world. Sevruk and Klemm (1989) analysed their global distribution and showed that the most widely used was the German Hellmann gauge, with 30 080 in use in 30 countries. The Chinese gauge was second, with 19 676 used in three countries, and the English Mark 2 and Snowdon third, with 17 856 operated in 29 countries, all three types totalling 67 612 and accounting for about half the world's (manual) raingauges. The Hellmann gauge has a funnel of 200 cm^2 or 15.96 cm diameter (6.28 inches) while the Chinese gauge has a collecting area of 314 cm^2 (diameter 20 cm or 7.9 inches). Like the Hellmann gauge, the Chinese gauge is made of galvanised iron.

The remainder of the world's most commonly used gauges are as follows:

Russia	13 620 gauges in seven countries, 200 cm^2, galvanised iron
USA	11 342 gauges in six countries, 324 cm^2 (8 inches), copper
India	10 975 gauges just in India, 200 cm^2, fibreglass
Australia	7639 gauges in three countries, 324 cm^2 (8 inches), galvanised iron
Brazil	6950 gauges just in Brazil, 400 cm^2, stainless steel
France	4876 gauges in 23 countries, 400 cm^2, galvanised or fibreglass
Total	55 402

The grand total of all these is 123 014. Other gauges in use elsewhere will bring the worldwide total up to Sevruk and Klemm's estimate of over 150 000.

No doubt this has changed since the estimate was made, and it probably missed gauges used by small organisations, but it gives a good idea of the situation. It does not include gauges that record the measurements automatically in any way.

Mechanical chart-recording gauges

Recording raingauges are used mostly to supply information on the time when rain starts and stops and to give an approximate indication of the rate of rainfall. It is usual to operate a manual gauge nearby to act as a reference and to give precise totals. Mechanical recording raingauges are of two main types – those that move a pen across a paper chart through the movement of a float and those that use a balance to move the pen.

Float-operated recorders

A typical chart-recording raingauge is the tilting siphon gauge designed by Dines (1920) (Fig. 8.2). The rain from the funnel is fed into a cylinder containing a float which rises as water enters, moving a pen up a paper chart. When the float approaches the top of the cylinder it releases a catch that causes the cylinder to tip over to one side. This action causes the siphon tube to be suddenly flooded with water, kick-starting the siphoning process and thereby preventing 'dribbling'.

Figure 8.2 To the left is the complete Dines tilting siphon recording raingauge with the paper chart visible through the window. To the right, the cover and funnel have been removed to show the mechanism. Collected water enters the container through the elongated funnel, raising a float which moves a pen up the chart. The container is balanced on a knife edge and when the container is full a catch is released that allows the container to tilt, thereby starting siphoning abruptly and precisely. When empty, a counterbalance returns the cylinder to the upright position to restart the cycle.

When nearly empty, and while still siphoning, the cylinder tips back to its vertical position ready to repeat the process.

The natural siphon recorder, designed by Negretti and Zambra, is similar to the tilting siphon recorder and predates it by 20 years. The siphon, however, consists of two coaxial tubes to the side of the float cylinder, constructed so that the annular gap between the inner and outer tubes is small. When the water reaches the top of the outer tube, capillary action ensures that siphoning starts decisively. After siphoning, and once air gets to the top of the tube, siphoning abruptly stops. The simple construction of this instrument means that it is cheaper and there is less to go wrong than there is with the tilting siphon gauge. It does not, however, continue to collect and save water during emptying, as the Dines does, this water being lost.

The Hellmann siphon gauge, designed in Germany in 1897, uses a siphon with a long tube that ensures clear-cut action. It became widely used in central Europe and, along with the Dines and Negretti recording gauges, probably accounts for most of this style of gauge still in use today.

Weight-operated recording gauges

Mechanical weighing raingauges operate by recording the weight of precipitation as it accumulates in a container, either by suspending the container on a spring or by

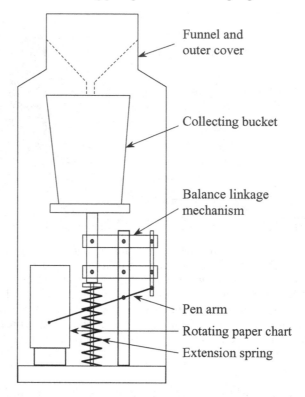

Figure 8.3 An alternative to a float mechanism is to weigh the collected water, and several mechanical weighing gauges have been designed, the general principle being illustrated in the figure. Most weighing gauges siphon the water out periodically so that they can operate continuously.

placing it on the arm of a balance (Fig. 8.3). In both designs the weight of the water forces the container downwards, the movement being magnified by lever linkages to move a pen.

Osler made one of the first weighing raingauges, in the UK in 1837, in which the container was counterbalanced by a weight on the opposite side of the fulcrum. A rod, connected to the weighted side of the balance, moved a pencil over a paper chart. The receiver had a siphon that emptied every 0.25 inches of rain. Two dozen or so different types of weighing recorder have been made in the period since 1837, some being emptied by hand, some siphoning the water out, while others used containers that tipped over when full, so emptying themselves.

The advantage of periodic emptying is not just that operation can be continuous but that the balance mechanism can be made more accurate because it does not have to work over a wide range of weight. For example, to measure 300 mm of rain to 0.1 mm requires a balance able to measure to 0.033% accuracy. But a mechanical balance in the field can only achieve about 0.5%, which would give a resolution of 1.5 mm of rain in 300 mm. To measure to 0.1 mm with an accuracy of 0.5% requires the collector to empty after each 20 mm of rain has fallen, which is perfectly feasible. In the case of float-operated recorders the same problem exists, but all modern float recorders siphon periodically, in such a way that the recorder needs to detect changes of 0.1 mm over a range of 5 mm, a resolution of only 1 in 50 (in the temperate models), or 0.1 mm over 25 mm, a resolution of 1 in 250 or 0.4% (in the tropical models). For anything other than approximate measurements, therefore, a weighing gauge, like a float gauge, must empty itself regularly.

Electrical raingauges

Tipping-bucket raingauges

Tipping-bucket gauges are by far the most common type of automatic raingauge in use today. Because of this we need to look more closely at them than we have at the older mechanical chart instruments.

Sir Christopher Wren's design (Chapter 7, Fig. 7.2) was single sided – it filled with water, tipped over and fell back again. In 1829 Crossley was the first to use a double-sided tipping bucket. Other designs followed throughout the nineteenth century in Europe and America and they proliferated in the twentieth century.

Modern designs are all symmetrical, although the shapes of the buckets vary from make to make (Fig. 8.4). The size of bucket is important, for it is undesirable to attempt to make a bucket tip with very small amounts of water because the errors (described below) increase. Ten millilitres or more is an advisable minimum, and so to measure 0.2 mm of rainfall reliably requires a funnel of at least 500 cm². Gauges are made with buckets that tip at intervals of 0.1, 0.2, 0.25, 0.5 and 1 mm.

The tipping of the bucket moves a magnet past a magnetic reed switch, giving a brief contact closure which would normally be recorded today by a data logger (see later), either built into the gauge as a single-channel recorder or as part of a multisensor system such as an AWS, which will include a multichannel logger. The contact closures of a gauge can be logged either as the total number of tips during a period or as the time and date of each tip, thereby giving an indication of intensity and time distribution of the rainfall.

The advantages of the tipping bucket are that it is simple, there is little to go wrong, it consumes no power and can be used with a variety of recorders. Its weaknesses are relatively few, but they can be a problem (Hanna 1995), as follows.

The record provided by a tipping bucket is not continuous, although if the tips are every 0.1 mm of rain this is close to being so. A UK Meteorological Office design overcame this difficulty by weighing the bucket as well as counting its tips, giving an indication of the increasing amount of water on a continuous basis between tips

Figure 8.4 Today, the most usual way to measure the water collected by a raingauge is by tipping bucket. If the collector is 500 cm^2 in area, 0.2 mm of rain produces 10 ml of water. As the bucket tips, a magnet in the moving arm momentarily closes a reed switch in the fixed outer arm, producing pulses that can be recorded, most usually by a data logger.

(Hewston and Sweet 1989, Whittaker 1991). It operates by suspending the bucket on wires, the natural resonant frequency of which increases as the weight increases, the signal being detected magnetically.

When a bucket has received the correct amount of water and starts to tip, water may continue to enter the 'full' side (if the rainfall is above a threshold rate) until it is in the level position, this rain being 'in excess' and thus lost, although the error is only small and only occurs at the higher rainfall rates. However, if the gauge is calibrated by passing water into it at controlled rates, this error can be calculated for different rain intensities and automatic correction made for it during data processing (Calder and Kidd 1978). (The WMO recently tested 19 gauges at rates of up to 300 millimetres per hour.) Such 'dynamic calibration' is preferable, in any case, to using a burette to pass water into the bucket until it tips once, since an average calibration for 50 tips will be more precise; drop-by-drop calibration is also difficult to do precisely. A carefully measured volume, of around 1 litre, is passed slowly into the gauge and the number of tips counted over a period of perhaps 60 minutes, the bucket being set to tip for the required amount by adjusting the pillars on which it rests. Since this can be a long process – of repeated adjustment and test – it can be preferable to set the bucket close to the desired sensitivity, of say 0.2 mm, and to measure exactly what it is by the passing of 1 litre of water. Thus a gauge may tip for each 0.19 mm, this being used as a bucket calibration factor.

Some water inevitably sticks to the bucket after it tips, and this is lost if the bucket does not tip again until after it has evaporated. If it does tip again, slightly more water will be required to tip the second time because the wet side is now slightly heavier, but on third and subsequent tips, balance is restored. If the bucket is calibrated, as above, by passing a litre of water through the gauge, the wetting effect is automatically compensated for and the error only occurs in the event of one isolated tip. The bucket's surface will change with use, generally accumulating a thin film of oxide, if made of metal, or of adhering dust, and the amount of water adhering to its surface may thus increase or decrease with time. There is, therefore, a case for calibration checks some months after installation. This is one reason why tipping buckets should be as large as possible, since in the case of small buckets the amount of water adhering is a larger percentage of their total capacity and so the error will be greater. To minimise the adherence of water, some tipping buckets have drip-wires at the point where the water leaves the bucket, although this only reduces the problem at that one point.

Because rain rarely ceases just as a bucket tips, some water is usually left standing in the bucket until it next rains. If this is within a few hours, there is not much loss, but if rain does not occur again for some time all the water left standing in the bucket may evaporate. While this can amount to one full tip, on average over a long period it will average out at half a tip. Depending on the type of rain experienced, this error can be large or small, light showers spaced by a day or so producing larger errors than long periods of heavy rain. This is another reason for having buckets that tip for small increments of rain (say 0.1 rather than 0.5 mm), for less will then be lost by evaporation.

Since the bucket must tip in response to a very small change in the amount of water accumulating in it (one drop), friction in the pivot or any variation of friction over time will affect performance. This is yet another reason for using as large a collector as possible. The type of bearing varies from make to make of gauge and includes ball races, knife edges, rolling systems and wires under tension. Each performs differently.

Electronic weighing raingauge

By replacing the pen of the mechanical weighing raingauge by a mechanism that turns a potentiometer, an electrical output is obtained that can be logged. In electronic versions, the water is weighed by a straingauge load cell. Because electronic weighing is more precise than can be achieved in the field with mechanical springs and levers, greater accuracy is possible. One commercial raingauge specification claims that increments of 0.01 mm of rain in a total of 250 mm – the capacity of its non-emptying container – can be sensed. Another model in this series has a capacity equivalent to 1000 mm of rain.

An advantage of the weighing gauge is its ability to measure the weight on a frequent basis (throughout rainfall), allowing the intensity of the rain to be measured in finer detail than by a tipping bucket. For long-term unattended operation, some means of emptying the weighed container automatically, usually by siphoning, is again necessary.

Capacitance raingauges

The collected water can also be measured by storing it in a cylinder containing two electrodes which act as the plates of a capacitor, the water acting as the dielectric between them (Fig. 8.5). The dielectric constant of water is around 80 and that of air 1. By including the capacitor in a tuned circuit it is possible to measure the depth of water. This can be sampled as often as required. As with other gauges that store the water, it must siphon it out periodically. This type of gauge is being used on buoys in the equatorial Pacific and Atlantic oceans in the Tropical Atmosphere and Ocean (TAO) project (see Chapter 14).

Figure 8.5 In this capacitance-based raingauge, the collected water is measured
by storing it in a deep cylinder which contains two electrodes acting as the plates
of a capacitor, the water between the plates performing the role of dielectric. By
including the capacitor in a tuned circuit, the depth of water can be measured. As
with other gauges that accumulate the water, it must be emptied periodically.

Drop-counting gauges

A few specialised gauges have been developed that convert the collected water
into drops of fixed size which are then counted (Norbury and White 1971). This
gives a good measure of the instantaneous rainfall rate as well as allowing totals
to be calculated. The weakness of the method is that the drop size can vary due

to changes in surface tension, caused by impurities in the water and temperature change. Norbury and White estimate that the drop volume can vary by 10% through these causes. While the method is useful for estimating second-by-second intensity (which is what it was designed for), 10% is not good enough for accurate totals. There is, however, the possibility that, combined with a tipping bucket (such that the water passes first through a drop counter and then into the bucket), a gauge could be produced that gives a good indication of both intensity and total rain; but this is a matter for future R&D.

Raingauge errors

Unfortunately it is not possible simply to put a raingauge outside and expect it to give an accurate measure of precipitation. There are many errors that frustrate this simple ideal and these need to be looked at next, along with possible preventative measures.

Size of the collector

In the 1860s, Colonel M. F. Ward did experiments in Wiltshire to compare the performance of rain collectors of different diameters, from 1 inch to 2 feet (2.54 to 60 cm) as well as square funnels. Catch was found to be independent of size, certainly above 4 inches. This was confirmed by tests by the Reverend Griffith, also in England in the 1860s, during which he operated 42 types of gauge (Kurtyka 1953, Reynolds 1965). Although one set of tests in the USA suggested that small gauges caught more than larger ones, it is probably safe to say that the diameter of the funnel does not matter provided it is bigger than about 10 cm.

The use to which the gauge is put in part decides the best size of the collector. For daily manual reading, too little water makes for inaccuracy, while too much needs a large container to hold it and makes measurement inconvenient. The UK Meteorological Office at one time preferred an 8-inch gauge but later changed to Symons's 5 inches, which is what is used today for manual, daily observations. (My opinion is that a 5-inch gauge catches too little water for convenient or accurate measurement.) If the gauge is of the mechanical chart-recording type, there must be sufficient water to overcome friction in the levers and to move the pen across the paper chart positively. In the case of a tipping-bucket gauge it is necessary to have a funnel big enough to collect at least 10 ml per tip, very small buckets tending to have higher errors. Because the water measured by recording gauges is not usually stored, there is no collection problem and so the bigger the collector, within reason, the better.

Evaporative loss

Traditionally, copper has been used to make raingauge funnels (in the UK), although today anodised aluminium, stainless steel, galvanised iron, fibreglass, bronze or plastic are all used. What is important about the material is how well it allows the water falling in the funnel to run off and be collected or measured; materials do not all behave in the same way in this respect.

Water first wets the surfaces of the funnel and the tubes leading to the bottle or measuring mechanism and only after this does it start to flow. When the rain stops, the water left adhering to the surfaces does not drain into the bottle or bucket but evaporates. In the case of a 5-inch UK manual gauge, the amount lost is about 0.2 mm per rain event. There can also be some loss from within the storage bottle, although this can be kept negligible by minimising ventilation and by housing the bottle below ground to keep it cool. A wire mesh within the funnel, acting as a filter to keep debris out, holds water and thereby increases evaporative loss. Any debris lying in the funnel will also hold water, which will thus be lost. If rain falls as intermittent light showers, loss by evaporation can amount to a significant proportion of the total catch, since it is cumulative. In the case of occasional heavy rain, however, the proportion is less.

Outsplash

The UK Meteorological Office 5-inch diameter gauge (Fig. 8.1) has vertical sides 4.5 inches (11.4 cm) deep and a steeply sloping funnel. This combination of depth and angle prevents virtually any outsplash of large drops or the rebound of hailstones. Many gauges are to this pattern, but others have less deep sides above the funnel and this will result in the loss of catch through outsplash in heavy rain. The disadvantage of deep sides, however, is that they increase the area to be wetted and so increase evaporative loss. Also deeper sides result in a tall cylindrical shape and this increases wind-induced errors (see below).

Levelling

I have seen many a raingauge leaning over at an angle. It may never have been installed correctly in the first place, or the ground may have settled after installation, or it may have been knocked by a lawnmower. Whatever the cause, the gauge will read wrongly. An error of about 1% occurs for each 1 degree of tilt according to Kurtyka (1953).

There is, however, an exception to this rule when installing a network of gauges in a mountainous area with extensive variation in slope of the ground for the purpose of measuring the areal average precipitation. In this technique the gauge orifice is angled to match that of the ground wherever it is placed.

Siting of gauges

A raingauge should ideally be installed no nearer to an object than four times the height of the object, although twice the height is acceptable. This applies not only to trees and buildings but also to other instruments such as temperature screens. Nevertheless the number of gauges that are not so exposed is surprisingly high and I have seen many, even at supposedly good sites, far too close to objects. Some sheltering may be beneficial, however, provided it is kept to the distance specified, because it lessens the adverse effects of wind. In a forested area, one solution is to site the gauge in a clearing, provided the clearing is big enough to meet the above criteria of distance. If this is not possible, gauges are sometimes installed on a mast just above the forest canopy.

The site itself must also be chosen to be representative of the surrounding area. More details of this task are given in Chapter 12 under the heading 'From point measurements to areal estimates'.

The effects of wind

From the last chapter it will be recalled that in the early nineteenth century there began to grow a dawning realisation that wind somehow caused raingauges to under-catch. Starting with the field tests of William Heberden in London, and the ideas and work of Henry Boase, William Jevons and George Symons, it was found over the course of 100 years that the problem was due to how the gauge interfered with the flow of wind. As soon as this was appreciated, methods were developed to combat it, and these are considered next. But even today few of the developments described below are used in practice at most raingauge sites. As a consequence of this, most of the world's raingauges under-catch by 5–50% depending on the design of the gauge and wind and rain conditions.

Wind shields

An early attempt to lessen the wind problem was made in the USA by Nipher (1878), who surrounded the gauge with an inverted trumpet-shaped shield that deflected the wind downwards (Fig. 8.6a). Such a shield can, however, become filled with snow and in 1910 Billwiller (in Germany) cut the bottom out of the Nipher shield to allow snow to pass through. He also removed the outer horizontal collar, although this was later shown to be an important part of the Nipher design. Also in the USA,

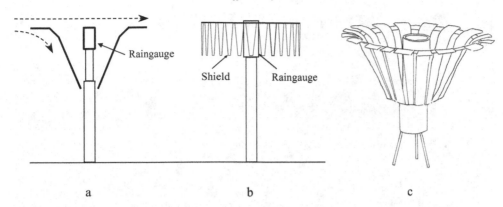

Figure 8.6 The Nipher shield (a) reduces wind effects by deflecting the wind downwards. However, it can collect snow, and to prevent this Billwiller modified it by removing the bottom part of the sides and the flat collar. The Alter shield (b) avoids the snow problem by making the sides out of leaves that swing inwards on the windward side in strong wind, simulating the shape of the Nipher shield. The Tretyakov screen (c) has the profile of the Alter and the individual (but fixed) leaves of the Alter, with gaps between to allow snow to pass.

Cecil Alter (1937) designed a shield consisting of loosely hanging hinged strips of metal which swung inwards on the windward side of the gauge, forming a conical shape similar to the Nipher profile while preventing the accumulation of snow (Fig. 8.6b). The Russian Tretyakov screen (Fig. 8.6c) is in effect a combination of the Alter and Nipher screens but with fixed plates.

The turf wall

From 1926 to 1933 Huddleston carried out extensive field tests at an upland site of 726 feet (221 m) altitude near to Penrith in the English Lake District to find the best raingauge exposure for mountainous and windy conditions. He compared the catch of ten UK Met Office 5-inch gauges exposed normally, in pits, surrounded by palisades and wrought iron fences as well as by wire-netting surrounds and turf walls of different sizes and heights. One also had a Nipher screen round it. After seven years of tests Huddleston (1933) concluded that the turf wall was best, stating that the height of the turf wall must be the same as the height of the gauge (Fig. 8.7).

The pit gauge

Although as long ago as 1822 Henry Boase had suggested that exposing a raingauge with its orifice at ground level would avoid the wind problem, no action was taken

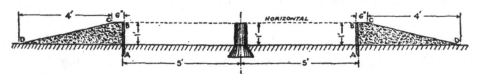

SECTION OF STANDARD CIRCULAR TURF-WALL FOR RAIN-GAUGES IN WINDY SITUATIONS.

A–B MINIATURE WOODEN FENCE.
B–C LEVEL PORTION OF WALL.
C–D SLOPING PORTION OF WALL, GRADIENT 1 IN 4.

Figure 8.7 Huddleston's turf wall produces, in effect, a slight hollow in the ground to shield the gauge. Top figure: a turf wall at the Institute of Hydrology in Walling-ford with a 5-inch manual gauge. It is surrounded by a variety of other gauges. In the foreground is a tipping-bucket gauge, in the middle distance is a Dines chart-recording gauge, to its left is a white experimental (in the 1970s) fibreglass tipping-bucket low-cost gauge. Above the Dines in the far distance is an exper-imental (1970s) aerodynamic gauge. On the far right is a 5-inch gauge exposed in the standard way at 1 ft height (30 cm). An anemometer is seen to the right of the Dines, at gauge height. In the far distance, centre right is, just visible, a 5-inch gauge exposed in a pit. This site was used for much of the experimental aerodynamic work carried out by Rodda in the 1960s and 1970s, described in this chapter. The bottom figure is Huddleston's drawing (1933).

until twenty years later when Stevenson (1842) buried a raingauge with its rim just above ground level, surrounded by a mat of bristles to prevent insplash. But the idea was still not adopted. Ninety years later in Germany, Koschmieder (1934) designed a pit gauge (Fig. 8.8), but again the idea was not adopted even though it largely eliminated the wind error. Thirty more years past before Rodda (1967a, 1967b)

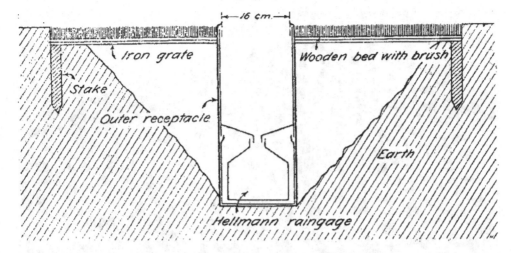

Figure 8.8 The Koschmieder pit places the gauge with its orifice at ground level surrounded by a brush mat to minimise insplash (from Koschmieder 1934).

reintroduced the idea at the Institute of Hydrology (now the Centre for Ecology and Hydrology) in Wallingford, in this case to ensure that an accurate measurement of precipitation input could be obtained for a long-term, detailed water balance experiment in central Wales. In this design (Fig. 8.9) a grating surrounds the gauge in place of the brush matting of Stevenson and Koschmieder to prevent insplash and to minimise eddies within the pit. Pits, however, have their problems since they can become snow- or sand-filled (as can the turf wall), while at unattended sites they can become overgrown with vegetation. In rocky or waterlogged ground they may be impracticable. Nevertheless, if possible, *the best way of exposing any raingauge is in a pit*, and this design was adopted by the World Meteorological Organization (Sevruk and Hamon 1984, WMO 1996). Pit gauge design was also recently the topic of a European Standards Working Group on raingauges (CEN/TC 318, WG 5, of which I was Convenor). A probable reason for pits not being widely adopted is because of the increased cost and work involved, but also through a certain amount of ignorance of the wind problem that still persists. In other cases it may just be that an approximate measure is sufficient.

Aerodynamic gauges

Another way of combating the effects of wind on raingauge catch is to design the profile of the gauge so as to present as little obstruction to the air flow as possible. Most gauges are cylindrical and these cause the maximum interference to the passage of the wind. Robinson and Rodda (1969) tested several raingauge designs in a wind tunnel, observing the flow of air with smoke trails. Field tests

Figure 8.9 Top: the Rodda pit grating, now the WMO standard. Lower three figures: Rodda tested several grating designs and found them all to be equally effective – so they can be simple and inexpensive. They share with the turf wall, however, the problem that they can fill with snow, with blowing sand and with growing vegetation, and they may be impossible in rocky terrain or waterlogged ground.

were also carried out to compare the catch of the gauges, including a funnel-shaped gauge of 5 inches (12.7 cm) diameter with a half-angle of 45 degrees, with no vertical sides above the funnel. This was the gauge which caught most rain except when it was heavy, when outsplash was suspected of causing losses.

Adopting a mathematical modelling approach, Folland (1988) analysed the flow of air over a cylindrical gauge and proposed a 'first-guess' new design, which he called the *flat champagne-glass* raingauge, with a diameter of about 25 cm and a half-angle of 35 degrees. While Folland was undertaking his theoretical work, I was also experimenting, independently, using the findings of Rodda and Robinson. The designs that emerged from these developments were similar to those of Folland's (Strangeways 1984). Field tests on two such designs (Hughes *et al.* 1993) confirmed that an improvement in catch was obtained with cone-shaped collectors, but that they did not give results as good as those from pit gauges, nor could the increased catch be unquestionably related exclusively to improved aerodynamics.

Over the last ten years I have carried out further and more detailed field tests, in particular over the four-year period from 2000 to 2004, in which gauges constructed to Folland's suggested design were tested alongside other collectors of modified and different profiles (Fig. 8.10). The results of this work have involved a re-valuation of the processes involved in wind flow over, around and within raingauges (Strangeways 2004a). In one series of tests, video records were made of the movement of a light nylon thread tracer and this demonstrated that there was a powerful circulating air current within the gauges. It had been traditionally assumed that the lift of the airflow over gauges was due simply to the presence of the gauge, but the tests suggest that the internal circulation is probably just as significant a contributor to the wind patterns (Fig. 8.11). In addition, the tests also highlighted what has tended to be ignored, that is that airflow in the real world is turbulent, not laminar as in a wind tunnel, and so the wind strikes gauges at many, and rapidly changing, angles and speed. Combined, these processes complicate the design (and mathematical modelling) of aerodynamic gauges.

Field intercomparisons and wind-tunnel tests

According to Sevruk and Klemm (1989) there are at least 50 different types of traditional manual raingauge in common use worldwide, all differing in size, shape, material, colour and the height at which they are exposed above the ground (from 0.2 to 2.0 metres). The number is greater if the newer automatic gauges are included. Because each gauge has its own characteristic errors, accurate intercomparison of rainfall data from around the world is not possible. To address this problem, the WMO organised field intercomparisons of the major national gauges of the world, with pit gauges as the reference, at 60 sites in 22 countries (Sevruk and Hamon 1984). However, such intercomparisons lump many rain events together over a period and can only give a rough correction. Windspeed and raindrop size are the two key factors in how different raingauges behave; if hour-by-hour measurements of these and of the rain are also available it becomes possible to correct readings with higher precision, but such are rarely available.

Sevruk (1996) has gone on to develop mathematical simulation techniques of the wind-induced error, using a three-dimensional numerical procedure. First the airflow around the gauge is simulated and then the trajectories of different drop sizes are computed in the simulated flow fields. The result is a formula giving

a

b

c

d

e

f

g

Figure 8.10 The gauges illustrated here are some of a series which I tested at a site near to Wallingford over four years from the year 2000 to evaluate the efficiency of various aerodynamic profiles (Strangeways 2004a): (a) shows two collectors to the Folland design (see text), at different heights; (b) is my modification to the Folland gauge, designed to reduce outsplash; (c) is (in the distance) a Folland gauge, a Folland gauge surrounded by a cylinder, a funnel-shaped gauge and (nearest) a 5-inch Mark 2 UK manual gauge as reference; (d) and (e) are gauges I designed while at the Institute of Hydrology in the 1980s (Hughes *et al.* 1993); (f) and (g) are gauges as seen in (b) but in pits at ground level for comparison.

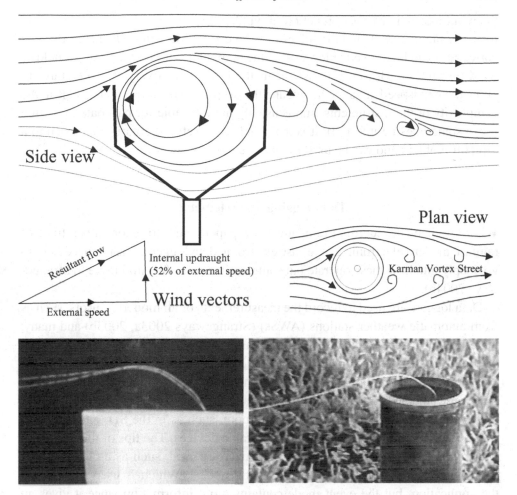

Figure 8.11 Field tests (Strangeways 2004a) using video records of light threads, tracing the wind movement around gauges, showed that there is a strong circulating current within the collecting funnels. The upward and outward movement on the windward side is around 50% of the external horizontal windspeed outside the gauge. The resultant flow over the top of the gauge is consequently affected not only by the presence of the gauge itself but also by the strong upward current out of the gauge. This effect has not been studied before when attempting to minimise aerodynamic losses. It occurs in all raingauges whatever their shape and size (lower photographs). For clarity, the drawing here shows the flow as perfect straight lines, although in reality it is very turbulent. Despite the external turbulence, the internal circulation is more consistent and laminar, being confined by the internal profile of the gauge.

the wind-induced error as a function of windspeed, rate of rainfall and drop size, for each raingauge. This is relatively new work and has the advantage that the corrections can be derived at low cost in the computer without the need for field or wind-tunnel tests. However, to use the method most effectively, it is preferable to know the windspeed at gauge height and the drop size spectrum, which adds to the complexity of the instruments, and is usually not available for past data (although it can sometimes be inferred). It is nevertheless a useful alternative approach to a difficult problem and can be used to improve past data.

Data logging and telemetry

While mechanical recording raingauges use paper charts to record a pen trace of rainfall, modern electronic gauges use electronic data loggers. It is not necessary to go deeply into how these operate but some outline is useful (find more in Strange-ways 2004b).

Data loggers are used to record the measurements of all modern sensors, such as from automatic weather stations (AWSs) (Strangeways 2003a, 2003b) and many, therefore, have to accept and log several channels. Loggers for use with raingauges on their own, however, can be simpler single-channel devices.

Whereas a paper chart is the recording medium for mechanical gauges, a data logger uses memory chips, usually RAM (random access memory) with capacities in the gigabyte range, and no doubt soon more, such as in the SD, XD and other cards used in digital cameras and other portable devices. The tips of the raingauge bucket can be recorded either as the total over a period, such as an hour or day, or the time and date of each tip (the 'event mode'). Which is used depends on the application, but the event mode contains more information since it gives an indication of the time distribution of the rain, and so its intensity.

While a paper chart has to be read manually and converted to a table of figures, the data stored in RAM in a logger can be automatically downloaded into a laptop computer on site (Fig. 8.12). The logger power supply can take many forms, from solar-charged lead/acid-gel batteries to lithium, which can power a logger for up to five years. And battery technology is advancing rapidly thanks to mobile phones and laptops. Data loggers free instruments from the need for human operators and so greatly broaden the geographical distribution of networks.

Multi-channel loggers have provision to process the various different sensor signals, mostly analogue, into a standard voltage, such as from a temperature sensor. These voltages are then switched in sequence to an analogue-to-digital converter (ADC) and thence to storage in memory. Raingauges that weigh the water, or that measure its capacitance, produce analogue signals, unlike the digital form of a tipping bucket, and so a logger with an ADC may be required. However, such

Figure 8.12 At this Papua New Guinea site, records from the tipping bucket raingauge (top) are being downloaded from a data logger (left) to a laptop.

raingauges may produce outputs that simulate tipping-bucket pulses or they may contain their own ADC.

While loggers are a great advance on paper charts, the ability to telemeter the measurements from a remote site to a distant base allows instruments to be operated unattended virtually anywhere. To telemeter the measurements, the digital records produced by a logger are first converted to voice-frequency tones by a modem, these tones then being transmitted over a telephone, land-based radio or satellite link (Fig. 8.13). How this is done is complex and we need not explore the details here (but see Strangeways 2004c, 2005).

Non-collecting raingauges

All the gauges so far described collect the rain in a funnel and measure the water in some way. Other, less direct, methods have been developed, and we now take a brief look at them.

Figure 8.13   Being able to transmit observations from a remote site allows measurements to be received in near real time at a distant base. At this British Antarctic
Survey site on the Antarctic Peninsula, readings are being telemetered via Meteosat.
The electronic unit, known as a data collection platform, is housed in the box in the
centre of the picture along with the battery power supply. The antenna is fixed to
the mast on the right. An aerodynamic raingauge is near, left (see also Fig. 8.10e).
This was part of an experimental cold-regions automatic weather station (inset)
(Strangeways 1985).

Optical raingauges

Optical raingauges do not collect the water but measure it directly as it falls. They
have evolved as a spin-off from visibility-measurement instruments, raindrops and
snowflakes being detected by their effect on a horizontal beam of infrared light
about one metre long (Fig. 8.14). An infrared receiver detects the passage of the
particles through the beam, the resultant scintillation being analysed by algorithms.
It is a rate-of-precipitation sensor although, by integration, hourly and daily means
of intensity can be obtained and thus the total amount of precipitation. The type
of precipitation, whether rain, snow, a mix of the two or hail, can be detected by

Figure 8.14 At the Rutherford Appleton field site at Chilbolton, this optical rain-gauge is used to detect rain intensity and drop size in radio propagation research. Optical gauges detect the precipitation passing through a beam of IR light passing between the source (right) and the detector (left) by analysing the scintillation produced by the falling drops. The nature of the fluctuations gives an indication of the intensity and type of precipitation, through complex algorithms.

analysing the spectrum of the scintillation. This type of sensor also gives a yes/no indication of precipitation, making it a useful instrument at, say, an airport or an Antarctic base, where an observer can watch a screen and know whether it is raining or snowing and if so how intensely; this can be particularly useful at night. For this reason another name for the instrument is a *present weather detector*. They are also useful where a large proportion of the precipitation falls as snow, which is difficult or impossible to measure with normal raingauges (see next chapter), although the need to keep the optics ice-free means that power requirements can be high. The algorithms for analysing the snow signal will also be less certain than for rain due to the great variety and size of snowflakes compared with the predictable shape of raindrops.

For indicating instantaneous precipitation rates, an optical raingauge has advantages over the tipping bucket, but for routine long-term collection of precipitation totals its high cost (about £7000 to £8000 currently) makes the tipping-bucket raingauge, at a fraction of the cost, inevitably preferable. Also, because the optical measurement is an indirect one, the conversion to rainfall intensity from the scintillating signal via an algorithm will not be perfect. Manufacturers of optical gauges (there are very few) quote an accuracy of from ±1% to ±10% (of actual intensity) depending on intensity and the type of precipitation.

Precipitation detectors

Precipitation detectors sense whether rain or snow is falling, but not its intensity. Another name for them is *ombroscopes*. A small sensing surface of somewhere between 10 and 50 cm^2 inclined at an angle (to allow water to run off) detects the presence of water on it by measuring either the change in resistance or the change in capacitance between interleaving electrodes, owing to the presence of water. The plate is heated if the temperature falls below +5 °C, to detect snow. It is also heated when there is water on it, so as to dry it off quickly after rainfall or snowfall stops. Lower-power heating prevents fog, dew and frost from being interpreted as rain. Ombroscopes have a shorter history than raingauges, having been made only over the last 100 years. They find use particularly in operational situations where the occurrence and duration of precipitation needs to be known in real time but where the quantity is not important.

Disdrometers

Impact-sizing sensors, disdrometers or *distrometers* measure the individual rain-drops, and were developed initially for cloud physics in the mid twentieth century. But drop-size detectors were also developed for use on the ground to measure the energy contained in the impact of each raindrop as it hit a surface, for example for soil erosion studies. They have also become important in the development of weather radar systems, both ground- and satellite-based (see Chapters 10 and 11), by providing a ground-truth measurement of drop size distribution.

Prior to the development of electrical disdrometers, various manual methods were used to detect the drops, such as allowing them to fall on a slate, on a chemically treated filter paper (the drops leaving a stain), and in bowls of flour or plaster of Paris. The spreading out of the drops, however, was a problem, as was their splashing on the slate. Today, electronic sensors detect the impact of the drops by piezoelectric transducers, such as in the disdrometer developed by Johns Hopkins University (Fig. 8.15), or by magnetic induction as a light Styrofoam body moves up and down in response to the drops, as developed by Joss and Waldvogel (1967)

Another technique is based on the measurement of the small changes in the value of a capacitor, made of interleaving electrodes (similar to those of a precipitation detector), covered by a thin glass plate. This type of sensor detects the changes in capacitance as each drop lands and spreads out on the glass, modulating the frequency of an oscillator circuit (Genrich 1989). It is the sudden arrival of the drop that is detected, the water remaining on the plate afterwards presenting a DC component that is ignored.

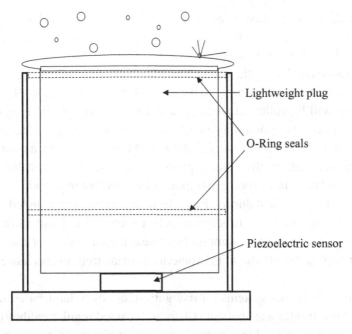

Figure 8.15 A number of different designs of disdrometer have been developed, the one illustrated being the JHU/APL instrument designed at Johns Hopkins University. The impact of the drops is transmitted via the lightweight plug to the piezoelectric sensor beneath, the resultant waveform being analysed to give drop size and rain intensity.

None of these are direct methods, however, and it is necessary to derive a reliable algorithm to convert the raw impact signal to the volume of the drop that produces it, and this will not be perfect. It is also inevitable that very small drops will not be easy to detect. (The WMO gives 0.3 mm as the smallest diameter reliably detectable and remarks that snow may not be detected at all. Thresholds will be different for each design and 0.3 mm is only a general guide.) With suitable algorithms, impact-sizing sensors can also measure rainfall totals.

Raingauges over the oceans

Most precipitation measurements are made on land, with very few from the oceans. All routine meteorological observations collected on ships today are made by Voluntary Observing Ships in the Voluntary Observing Fleet (VOF). There are over 6000 such ships and the *Marine Observer's Handbook* (Met Office 1995) describes these and their measurements in detail. But many factors reduce the effectiveness of a raingauge when operated on a ship (Hasse *et al.* 1997, Yuter and Parker 2001). Rolling and pitching requires the gauge to be kept horizontal by some form of gimbal, but they are expensive and not completely effective. The higher winds experienced at sea, together with the forward motion of the ship and the very open exposure, make the wind-induced errors even greater than those experienced on land, further worsened since the gauge has to be exposed amongst the ship's super-structure, which causes turbulence. To these problems must be added the possibility that sea spray will be collected along with the rain. There are no simple answers to these problems. One possibility is the use of aerodynamic collectors (Folland 1988, Hasse *et al.* 1997, Strangeways 2004a), but this can only be a partial solution since it does not address the other problems. Yet a further difficulty is that the rainfall total will refer not to one geographical location but to the whole area across which the ship has travelled during the collection period, compounded by the 'fair weather bias' – the tendency of ships to select routes that avoid storms, thereby not obtaining a representative sample. For these reasons none of the VOF ships measures precipitation, although it is sometimes measured on specialised research ships.

Large ships have radar systems for navigation, but even land-based radars used specifically for rainfall measurement (Chapter 10) need regular calibration against ground-truth raingauges. Ships' radars are susceptible to drift, which cannot be corrected in this way. A ship's radar is thus not a realistic option for precipitation measurement over the oceans.

On buoys, the exposure problems are perhaps slightly less problematic than for ships, because the buoy is at a fixed point and just a few metres above the surface. In the case of the buoys used in the Tropical Atmosphere and Ocean (TAO) project (see Chapter 14), the capacitive type of raingauge is used (see above). However, no gimbals are used since it is considered that while the swaying of the buoys will cause some error, the waves are normally sufficiently small to cause only minimal problems. This is something that a tipping-bucket raingauge could not cope with. The greatest source of error is considered to be the wind, but again this is felt to be sufficiently light across the equatorial Pacific and Atlantic to be tolerated. While the measurements will certainly contain errors, the measurements are valuable and far better than no measurements at all.

Techniques are also being developed (Quartly *et al.* 2002) for use on buoys to sense rainfall at sea by analysing the sound-signature of the air bubbles produced by the impact of raindrops. Its very indirectness means that high precision will be difficult to achieve, but it will present another useful approach. During develop-ment these methods will need precise ground-truth measurements from accurate raingauges to allow their algorithms to be refined.

As a result of these difficulties, there are very few rainfall data from either ship or buoy. Only with the coming of satellite remote sensing has this problem partly eased, as will be shown in Chapters 11 and 12.

Most rainfall data from the oceans are collected by gauges on islands. The main problem with islands as platforms is that their presence can modify the precipitation in their locality. But with careful site selection this can be minimised, and smaller islands will not have any effect on precipitation.

Metadata

If raingauge measurements are made for an extended period (decades) it is probable that conditions at the site where the raingauge is operated will change, for example trees may grow, or buildings may be erected, or the enclosure may even be moved to a nearby new location. A gauge may be damaged and have to be replaced, or a new type of gauge may replace an older one, and automatic gauges will need periodic recalibration. These changes affect the readings, sometimes slowly, sometimes suddenly. If the precipitation measurements are to be used for the detection of trends over time, these changes need to be allowed for. It is most desirable, therefore, that a record of any changes is kept, these being known as 'metadata'. However, it is rare that a good record is kept and this makes the detection of changes in precipitation difficult; more on this in Chapters 13 and 14.

The aerodynamics of gauges when they are used to collect snowfall are even more extreme than for rainfall, and the next chapter looks into this and at what can be done to lessen the quite considerable errors.

References

Alter, J. C. (1937). Shielded storage precipitation gauges. *Monthly Weather Review*, **65**, 262–265.

Calder, I. R. and Kidd, C. H. R. (1978). A note on the dynamic calibration of tipping bucket raingauges. *Journal of Hydrology*, **39**, 383–386.

Dines, W. H. (1920). The tilting raingauge: a new autographic instrument. *Meteorological Magazine*, **55**, 112–113.

Folland, C. K. (1988). Numerical models of the raingauge exposure problem, field experiments and an improved collector design. *Quarterly Journal of the Royal Meteorological Society*, **114**, 1485–1516.

Genrich, V. (1989). Introducing 'electronic impact sizing' as a new, cost-effective technique for the on-line evaluation of precipitation. In *Proceedings of the WMO/IAHS/ETH International Workshop on Precipitation Measurement*, St Moritz, pp. 211–216.

Hanna, E. (1995). How effective are tipping-bucket raingauges? A review. *Weather*, **50**, 336–342.

Hasse, L., Grossklaus, M., Uhlig, K. and Timm, P. (1997). A ship raingauge for use in high winds. *Journal of Atmospheric and Oceanic Technology*, **15**, 380–386.

Hewston, G. M. and Sweet, S. H. (1989). Trials use of weighing tipping-bucket raingauge. *Meteorological Magazine*, **118**, 132–134.

Huddleston, F. (1933). A summary of seven years' experiments with raingauge shields in exposed positions, 1926–1932 at Hutton John, Penrith. *British Rainfall*, **73**, 274–293.

Hughes, C., Strangeways, I. C. and Roberts, A. M. (1993). Field evaluation of two aerodynamic raingauges. *Weather*, **48**, 66–71.

Joss, J. and Waldvogel, A. (1967). Comments on 'Some observations on the Joss Waldvogel rainfall disdrometer'; reply by P. I. A. Kinnell. *Journal of Applied Meteorology*, **16**, 112–114.

Koschmieder, H. (1934). Methods and results of definite rain measurements III. Danzig Report (1). *Monthly Weather Review*, **62**, 5–7.

Kurtyka, J. C. (1953). *Precipitation Measurement Study*. Report of Investigation 20, State Water Survey, Illinois. Urbana, IL: Department of Registration and Education.

Met Office (1995). *The Marine Observer's Handbook*. London: Stationery Office.

Nipher, F. E. (1878). On the determination of the true rainfall in elevated gauges. *American Association for the Advancement of Science*, **27**, 103–108.

Norbury, J. R. and White, W. J. (1971). A rapid response rain gauge. *Journal of Physics E: Scientific Instruments*, **4**, 601–602.

Quartly, G. D., Guymer, K. G. and Birch, K. G. (2002). Back to basics: measuring rainfall at sea. Part 1 – in situ sensors. *Weather*, **57**, 315–320.

Reynolds, G. (1965). A history of raingauges. *Weather*, **20**, 106–114.

Robinson, A. C. and Rodda, J. C. (1969). Rain, wind and the aerodynamic characteristics of raingauges. *Meteorological Magazine*, **98**, 113–120.

Rodda, J. C. (1967a). The rainfall measurement problem. In *Proceedings of the Bern Assembly of IAHS*, pp. 215–231.

Rodda, J. C. (1967b). The systematic error in rainfall measurement. *Journal of the Institute of Water Engineers*, **21**, 173–179.

Sevruk, B. (1996). Adjustment of tipping-bucket precipitation gauge measurements. *Atmospheric Research*, **42**, 237–246.

Sevruk, B. and Hamon, W. R. (1984). *International Comparisons of National Precipitation Gauges with a Reference Pit Gauge*. WMO Instruments and Observing Methods, Report 17. WMO/TD, No. 38.

Sevruk, B. and Klemm, S. (1989). Types of standard precipitation gauges. In *WMO/IAHS/ETH International Workshop on Precipitation Measurement*, St Moritz.

Stevenson, T. (1842). On the defects of raingauges with descriptions of an improved form. *Edinburgh New Philosophical Journal*, **33**, 12–21.

Strangeways, I. C. (1984). Low cost hydrological data collection. In *Proceedings of the IAHS Symposium on Challenges in African Hydrology and Water Resources, Harare*. IAHS Publication 144, pp. 229–233.

Strangeways, I. C. (1985). The development of an automatic weather station for arctic regions. Unpublished Ph.D. thesis, University of Reading.

Strangeways, I. C. (2003a). Back to basics. The 'met. enclosure', Part 9(a): automatic weather stations. Temperature, humidity, barometric pressure and wind. *Weather*, **58**, 428–434.

Strangeways, I. C. (2003b). Back to basics. The 'met. enclosure', Part 9(b): automatic weather stations. Radiation, evaporation, precipitation, ocean buoys and cold regions. *Weather*, **58**, 466–471.

Strangeways, I. C. (2004a). Improving precipitation measurement. *International Journal of Climatology*, **24**, 1443–1460.

Strangeways, I. C. (2004b). Back to basics. The 'met. enclosure', Part 10: data loggers. *Weather*, **59**, 185–189.

Strangeways, I. C. (2004c). Back to basics. The 'met. enclosure', Part 11: telemetry by telephone and ground-based radio. *Weather*, **59**, 279–281.

Strangeways, I. C. (2005). Back to basics. The 'met. enclosure', Part 12: Telemetry by satellite. *Weather*, **60**, 45–49.

Whittaker, A. E. (1991). Precipitation rate measurement. *Weather*, **46**, 321–324.

World Meteorological Organization (1996). *Guide to Meteorological Instruments and Methods of Observation*, 6th edn. WMO, No. 8. Geneva: WMO.

Yuter, S. E. and Parker, W. S. (2001). Rainfall measurement on ship revisited: the 1997 PACS TEPPS cruise. *Journal of Applied Meteorology*, **40**, 1003–1018.

9

Measuring snow

Snowfall is much more difficult to measure than rainfall because the crystals are deflected by the wind to a greater extent due to their lightness and size, worsening the aerodynamic problems. Above a windspeed of a few metres per second, the use of raingauges for measuring snow is not feasible without some form of screen. Pit gauges are obviously not practicable for anything but very small falls, although a possible alternative is the use of aerodynamically shaped collectors (Fig. 8.10). After entering the funnel, internal eddies can also lift out dry snow that has been successfully caught (Fig. 8.11). This can be reduced by putting a cross-shaped vertical divider in the funnel, although this increases evaporative loss when the gauge is measuring rain.

As the windspeed increases so does the angle of fall of the snowflakes, becoming increasingly more horizontal, approaching the gauge at such a shallow angle that the orifice presents too thin an ellipse for them to enter.

After falling, dry snow can also be carried elsewhere by the wind as spindrift, which if caught by gauges cannot be differentiated from true snowfall (Fig. 9.1).

Measuring snow as it falls

Measuring snow with manual raingauges

If the snowfall does not overcap or bury the gauge and the wind is light, it is possible to use conventional manual raingauges to measure snow reasonably well. If the snow has not melted naturally at the time of reading, it must be melted in the funnel and the resultant liquid water measured in the usual way.

If amounts of snow are greater and the windspeed remains low, the funnel can be heated to melt the snow as it falls. This needs power, which can be either electrical or bottled gas. Alternatively the funnel can be removed so that the snow falls directly into an open container, the rim being retained to define the catch area. Without

Figure 9.1 Dry snow can blow like sand, as here on Deception Island off the Antarctic Peninsula and (inset) on the Cairngorm mountains in Scotland. If caught by raingauges, spindrift will be mistakenly interpreted as snowfall.

some means of melting the snow, however, even a gauge of this design will become full quickly. Some open gauges use antifreeze to melt the snow. But open gauges suffer from greater evaporative losses than those with funnels, although this can be prevented by floating a thin layer of light oil on the melted snow. But in high winds this is all to no avail; catching the correct amount of snow comes first.

Measuring snow with recording raingauges

If falls do not overcap the funnel or bury the gauge, natural melting is occurring and the wind is light, siphoning, weighing and tipping-bucket gauges will all measure snow satisfactorily, provided that their mechanisms do not freeze up; some gauges have internal heaters to prevent this. The timings of the record will indicate melting time, not falling time, even though the total may be correct. It is possible to heat a funnel to supply the necessary energy to melt the snow but this takes from 250 to 1000 watts depending on temperature and windspeed. If, however, the wind is above a few metres per second, the aerodynamic losses will be very great and the melting of the snow again becomes merely academic.

By removing the funnel completely, just as with a manual gauge, it is possible to collect the snow directly in a large open container, the build-up of snow being weighed mechanically or electronically. Simpler designs do not melt the snow and so are of limited capacity unless natural melting occurs; more complex types use antifreeze. A further refinement is that of emptying the container and automatically replenishing the antifreeze, retaining an anti-evaporation oil film by stopping siphoning in good time. Some users object to the emptying of antifreeze into the ground, due to environmental considerations. One design of recording snow gauge omits the oil, measuring the loss of weight due to evaporation as well as the gain through snowfall.

Measuring snow with an optical precipitation gauge

Optical gauges (see previous chapter) can also measure snow, and since they do not catch or melt it many of the problems experienced with raingauges are avoided. But while the water content of raindrops is precisely known from their size, thereby allowing a moderately precise estimate of rain intensity to be inferred from the scintillation signal, snowflakes vary widely in shape and density (Chapter 5), making the conversion from light signal to intensity of snowfall much less certain than for rain.

Wind shields

Because of the large wind errors in measuring snow, attempts have been made to shield raingauges when they are used to measure snow, the Alter shield being an early example (see previous chapter). The best measurement of snowfall using raingauges was deemed by the WMO to be that obtained by having bushes around the gauge cut to gauge height, or to site the gauge in a forest clearing. But bush screens are only practicable if the bushes themselves do not become full of, or covered by, snow. Also, because suitable forest clearings or bushes may well not be available at many cold sites, and as they will all be different, they were not selected by the WMO as the reference. Instead, the WMO designated the *octagonal, vertical, double-fence shield* as the intercomparison reference (DFIR) against which all snow gauges should be compared (Fig. 9.2).

A comparison of the DFIR was made with bush shields at Valdai in Russia, the bush exposure catching most snow, the DFIR 92% of this. All other gauges caught significantly less: for example, compared with the bush gauge, an unshielded 8-inch manual gauge caught 57% while an identical gauge with an Alter shield caught 75%. Methods have been developed to correct the DFIR shield to the bush readings (Yang *et al.* 1994) and these can then be compared with the catch of national gauges in an attempt to correct the national readings. This was done for 20 different gauge and shield combinations (Metcalfe and Goodison 1993). An example of the findings of the comparison was that the catch of Hellman-type gauges could be as

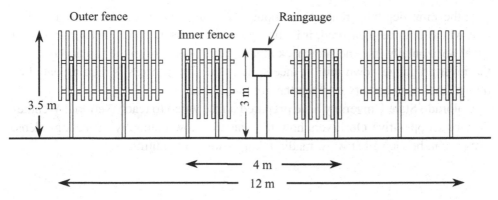

Figure 9.2 While wind effects can cause large errors in rainfall catch, the effect is much greater in the case of snow. The WMO recommends the use of the double-fence intercomparison reference (DFIR) as an international reference against which to compare the performance of different gauges and different exposure methods. But the double-fence snow shield is large, as the dimensions in the figure show, and so it is costly and not suited to widespread, general use, for example as a snow counterpart to a raingauge pit. It consists of two hexagonal fences, one within the other, surrounding the raingauge, which may also have an Alter or similar screen around it for good measure. In the figure, parts of the fence are omitted for clarity.

little as 25% of the corrected DFIR gauge reading in winds above 5 m s^{-1} (Gunther 1993).

Although correction procedures and field tests comparing national gauges with the DFIR are necessary, so as to be able to use the already existing data, the large size and expense of the double snow fence makes it impractical for general use. Since raingauges are, in so many ways, an unreliable means of measuring snow, alternatives have been sought.

Measurement of snow after it has fallen

Snow sections

After seven years of field tests 150 years ago, the only way that Huddleston could see of getting sensible readings was to take a 'cheese' of snow off the ground where it seemed to be of average depth, and to melt it. This still represents one of the best ways of measuring snowfall – but it needs an observer. After small falls of snow it is possible to collect a sample by inverting a raingauge funnel and pressing it through the snow until it meets the ground. The snow in the funnel is then melted and the water measured. If there is doubt about whether the chosen site is typical, several samples can be taken some distance apart, the measurements being recorded separately and the mean quoted.

If the snow depth is greater than about 15 cm and an area can be cleared next to that which is to be measured, it is then possible to insert a thin metal or wooden sheet horizontally into the snowpack at a suitable height from the top and to press the inverted gauge down until it meets the plate. The process is then repeated as many times as necessary to reach the ground.

Colorado State University have prepared a free video to teach National Weather Service Cooperative Observers how to measure snow manually in their gardens. This is can be viewed at www.madis-fsl.org/snow_video.html.

Snow depth

It is relatively easy to measure the depth of snow manually, graduated poles fixed in the ground simply being read at intervals. Automatic level sensors have also been developed, sensing the top surface of the snow by downward-looking ultrasonic echo systems. They typically measure the depth to about ± 2 cm, but falling snow and spindrift can interfere with the readings, spurious signals being received as noise from the flakes. The air temperature also has to be measured to compensate for the variation in the speed of sound with temperature.

However, the snow's density is unknown, varying considerably from place to place and, as it is the water-equivalent of the snow that is usually required, without knowledge of snow density this information cannot be deduced from the depth. Various rules of thumb are used, such as the one-tenth rule, but snow density can vary from one tenth to one fiftieth, an average being about one thirteenth. Taking test samples occasionally, to establish the density of the nearby snow, improves the estimate but is laborious and cannot be automated.

Snow weight

A way around not knowing the density of the snow is to measure the weight of the snowpack where it lies using 'snow pillows'. These are thin containers made of flexible sheets of stainless steel, rubber or plastic, about a metre square and a centimetre or so thick, with a capacity of around 50 litres of liquid. The liquid is usually a solution of antifreeze, the weight of snow being measured with an electronic pressure sensor or a float-operated level recorder in a standpipe.

However, the snowpack slowly changes with time and can become solid or form ice bridges so that gradually its full weight may no longer rest on the pillow. A larger pillow can reduce but not eliminate this problem. If the snow does not stay for long and is not deep, bridging and solidification may not have time to occur.

Gamma ray attenuation

If a collated beam of gamma rays passes through the snowpack, it becomes attenuated to an extent depending on the water content of the snow. This is an effective,
if expensive, technique but even though the gamma source is low level (typically
10 mCi of caesium 137), health concerns can cause problems. Consequently it is not
in wide use. Natural background radiation is also sometimes used in a similar way.

Measurement of snow as it melts

Melting snow is of concern to hydrologists and to river managers, so information
on melt rate and time is well worth having even if, due to evaporation and spindrift, it does not tell us exactly how much fell in the first place, or when it fell.
Snowmelt lysimeters collect and measure the meltwater flowing from the bottom
of the snowpack and take two forms – unenclosed and enclosed – the former being
much the commoner. In this design the collector has a shallow wall around it, while
in the latter the wall completely isolates the snow column from its surroundings all
the way up to the surface of the pack. In the latter design, the wall may be raised
slowly as the snowpack increases in depth, although this is not of course possible
at an unattended site.

Meltwater percolating down through the ice pack follows an indirect downwards
route, travelling laterally along the less permeable layers until it finds a way through
(Wankiewicz 1979) (Fig. 9.3). The lysimeter collects this flow, but as most lysimeters are small what they collect may not be entirely representative. The bigger the
lysimeter, therefore, the better, the collecting area varying in practice from less than
1 m^2 up to 100 m^2.

Meltwater flows mostly under the influence of gravity (Colbeck 1972, 1974),
capillary effects being negligible except at discontinuities (Wankiewicz 1979). The
base of the lysimeter is a discontinuity (Fig. 9.3), the layer at atmospheric pressure
causing a saturated layer of up to 2–3 cm depth to form (Kattelmann 1984), the
water being held in the spaces between the snow grains by capillary force. Once
formed, this saturated layer remains fairly constant. Because the saturated layer is
only a few centimetres deep it can be contained within a lysimeter wall provided
it is from 12 to 15 cm deep, thereby avoiding the problem. However, the presence
of this zone affects the timing of meltwater outflow from the lysimeter, since the
saturated zone has first to form – a process that can take up to several days initially.
Thereafter, the response to percolating water can be within hours or even minutes.
The problem can be largely overcome by using tension lysimeters, in which the base
of the lysimeter is a porous plate held at a negative pressure approximating to that
of the capillary pressure of the overlying snow, much reducing the response time.

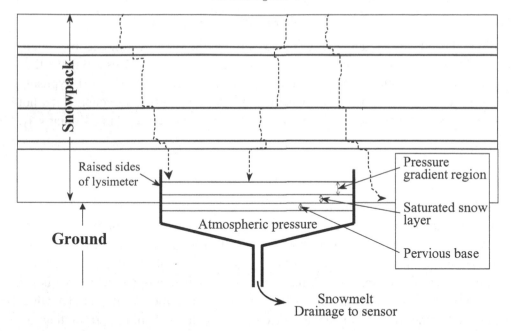

Figure 9.3 A snowpack may not be homogeneous but composed of layers of different permeability and crystal structure, causing the meltwater to percolate down
under gravity by diverse paths. To collect a representative sample of water, a
snowmelt lysimeter must be large enough to even out any such spatial variations.
Because the lysimeter is a discontinuity within the pack, its sides must be about
15 cm high to prevent snowmelt going round rather than into the collector due to
differences in tension causing pressure gradients. Once caught, the water can be
measured by a variety of means from manual to automatic using a tipping bucket.

The meltwater is finally piped to a suitable site and can be collected in a container
and read manually or more usually be measured automatically using a tipping
bucket.

There is no simple solution to snowfall measurement, not even by radar, which
is the topic of the next chapter.

References

Colbeck, S. C. (1972). A theory of water percolation in snow. *Journal of Glaciology*, **11**,
    369–385.
Colbeck, S. C. (1974). The capillary effects on water percolation in homogeneous snow.
    Journal of Glaciology, **13**, 85–97.
Gunther, T. (1993). German participation in the WMO solid precipitation
    intercomparison: final results. In *Proceedings of the Symposium on Precipitation and
    Evaporation*. Bratislava: Slovak Hydrometeorological Institute, vol. 1, pp. 93–102.
Kattelmann, S. C. (1984). Snowmelt lysimeters: design and use. In *Proceedings of the
    Western Snow Conference*. Fort Collins, CO: Colorado State University, pp. 68–79.

Metcalfe, J. R. and Goodison, B. E. (1993). Correction of Canadian winter precipitation data. In *Proceedings of the 8th Symposium of Meteorological Observations and Instrumentation*. Anaheim, CA: AMS, pp. 338–343.

Wankiewicz, A. (1979). A review of water movement in snow. In *Proceedings Modeling of Snow Cover Runoff*. Hanover, NH: US Army Corps of Engineers Cold Regions Research and Engineering Laboratory, pp. 222–252.

Yang, D., Sevruk, B., Elomaa, E., Golubev, B., Goodison, B. and Gunther, T. (1994). Wind-induced error in snow measurement: WMO intercomparison results. In *23. Internationale Tagung für Alpine Meteorologie. Annalen der Meteorologie*, **30**, 61–64.

10

Measuring precipitation with radar

Its short history

Radar was developed in the 1940s for the detection of aircraft, and the effects of precipitation on the received signal were originally seen as an inconvenient source of interference. However, even in the late 1940s it was recognised that radar could be used to measure precipitation. Kurtyka (1953) remarked that 'In the last five years, the necessity of adequate precipitation instruments to calibrate radar for precipitation measurement has pointed to the primitiveness of the present-day rain gauge.' He went on to say that 'In all likelihood, the rain gauge of the future may be radar, for even in its present developmental stage, radar measures rainfall more accurately than a network of one rain gauge per 200 square miles.' But radar has not replaced raingauges, and while it has advanced greatly over the years, it still relies on *in situ* data from telemetering raingauges to calibrate the system, although this need may eventually be overcome.

Principle of operation

One of the great advantages of radar is that it gives an areal estimate of precipitation rather than a single-point measurement, and the area covered is quite large, typically about 15 000 km^2 for each station. It also has the advantages of giving data in real time and of not needing anything to be installed in the area, or even access to the area.

In a weather radar system, the dish alternately transmits a pulse of microwaves and then switches to receive the returned (or more correctly 'scattered') pulse. The beam emitted is a cone of between 1 and 2 degrees divergence. From the return time of the pulse the range of the target can be calculated, while from the strength of the returned signal the reflectivity of the precipitation can be estimated. Any Doppler shift of the returned frequency establishes the velocity of the target (in the direction of the beam), while the polarisation of the signal gives information on the shape of the precipitation and its orientation.

The wavelengths used are generally 3 cm (X band or 10 GHz), 5 cm (C band or 6 GHz) and 10 cm (S band or 3 GHz), these wavelengths being scattered mostly by targets the size of precipitation particles. The 3- and 5-centimetre bands do not require a large dish, which keeps costs down and helps in portability, but these frequencies are attenuated when they pass through heavy rain. At 10 cm and longer, attenuation is much reduced, but the dish required increases to a diameter of 8 metres for a 1-degree beam. Nevertheless 'primary' land-based radars use this frequency, while on ships and aircraft 5 or 3 cm have to be used to minimise dish size. On satellites around 2 cm (K_u band or 15 GHz) is used with a phased array rather than a dish (see the Tropical Rainfall Measuring Mission satellite, Chapter 11, Fig. 11.6).

As the dish rotates through 360 degrees over a period of from 1 to 10 minutes it emits pulses, each pulse being of a length that spans a distance of about 100 m, giving reasonable distance resolution. But as the beam width is between 1 and 2 degrees, the width resolution is less and increases with distance from the antenna, being around 1 km near the radar but several kilometres at its outer ranges. Detection of precipitation extends as far as 200–400 km but quantitative measurements are limited to 100–200 km.

The rain, snow, graupel and hail detected by the radar are known as *distributed targets* because the radar beam is scattered by many discrete small elements, rather than reflected by one large one such as an aircraft. The precipitation particles are sufficiently small (diameter less than 0.1 of the wavelength) for Raleigh (or 'elastic') scattering theory to apply (scattering in which there is no change in frequency, in contrast to Raman (or inelastic) scattering, where the frequency changes upon scattering).

The volume illuminated by the beam is known as the *resolution volume*, this being set by the width of the beam and the length of the pulse. The precipitation targets are all moving relative to each other due to the different fall speeds of drops of various diameters, and through wind shear and turbulence, causing the amplitude and phase of the individual scattered pulses to fluctuate continuously. By taking an average of the returned signals over 0.01 to 0.1 seconds, the fluctuations are removed.

Estimates of the rainfall rate in millimetres per hour, R, are derived from the returned energy, or *radar reflectivity*, Z, using an empirical equation of the form:

$$Z = aR^b$$

Both a and b have many possible values, the most usual being $a = 200$, $b = 1.6$. The value of b is usually left at about this figure, while a may lie anywhere from 140 for drizzle, through 180 for widespread frontal rain, to 240 for heavy convective showers. These are figures for the UK, but in other situations they can be different, for example 500 for thunderstorms in an Alpine setting. These values are quoted to illustrate the magnitude of the variation of the reflected signal from different rain types and at different places; yet other values of *a* apply to snow.

Because falling precipitation follows the horizontal wind, the Doppler frequency shift allows windspeed to be measured to 0.1 m s^{-1} (when it is raining – although other scatterers can also be used). While it is the speed in line with the radar beam that is measured the other velocity components can be calculated if the wind direction is known.

Operational systems

A classic weather radar system which had been very thoroughly evaluated by a consortium comprising North West Water, the Meteorological Office, the Water Research Centre, the Department of the Environment and the Ministry of Agriculture, Fisheries and Food was the first unmanned weather radar project in the UK, installed on Hameldon Hill, about 30 km north of Manchester, in 1978. This brought within range the hills of north Wales, the southern Lake District and the Pennines and so gave the opportunity to evaluate a system in different terrains (Collier 1985). The performance and characteristics of this system are considered later.

Weather radar is now fully operational in the UK and currently there are 12 weather radars spread fairly evenly from the south Cornish coast to the Outer Hebrides, covering the whole country and beyond over the sea. There are also two in the Republic of Ireland and one in the Channel Islands (Fig. 10.1). In principle they are the same as that pioneered at Hameldon Hill, this being typical of many similar systems worldwide.

Sources of error

Uncertainty in converting from reflectivity to precipitation intensity is due to many factors, including the variation in drop-size spectrum from one event to another and the presence of hail and snow. In addition, melting snow increases the reflectivity, giving what is called the *bright band*, leading to overestimation of the rain intensity. Conversely, drizzle tends to be underestimated. That the factor *a* has so many different values depending on rainfall type and location underlines the uncertainty of conversion from returned signal to precipitation intensity.

A further cause of uncertainty is the inability to detect precipitation beneath the radar beam. A low beam elevation is necessary to detect rain as near to the ground as possible and to cover as long a range as possible, but hills interfere with too low

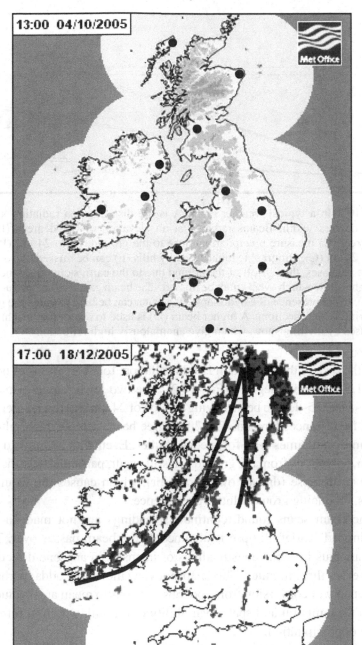

Figure 10.1 Areas of rainfall measured by the weather radars over the UK, the Channel Islands and the Republic of Ireland. The upper map shows a day of no rainfall, with the approximate locations of the radars shown by black discs. The lower map shows rainfall at a warm front (over the east coast of Scotland and England) and a cold front over the east coast of Ireland with the warm sector between. I have added lines marking the fronts. The images are much clearer when in colour but the areas of rainfall can be made out by comparing the two maps. Crown copyright 2005. Published by the Met Office.

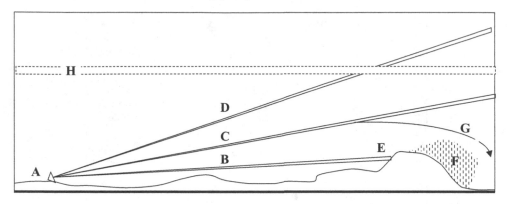

Figure 10.2 In a typical weather radar, A is the dish antenna radiating a beam of 1–2 degrees width. Beams radiated at an elevation of 0.5 degrees (B) and 1.5 degrees (C) measure precipitation close to the ground, up to 24 km (B), and beyond 24 km (C). Orographic rain caused by hills (F) can be missed. Also, as the distance increases, the fall-off of the ground due to the earth's curvature increases the height below which events may be missed. The beam can also be obstructed by hills (E). In the presence of a hydrolapse, the beam can be bent towards the ground (G), giving false reflections. A higher beam (D) is used to detect the 'bright band' (H) caused by melting snow, which give anomalously high reflections.

a beam. In the case of the Hameldon Hill installation, four elevations of beam were used: 0.5, 1.5, 2.5 and 4 degrees (Fig. 10.2). The two lower angles sense surface precipitation, the 0.5-degree beam having a range of 24 km and the 1.5-degree beam being used for distances beyond. The 2.5-degree beam detects the bright band. A higher beam is sometimes used for special studies. Even with an upward elevation of the beam, reflections from the ground can still occur, particularly in the presence of a strong hydrolapse (dry air over moist air), which causes the beam to bend downwards, producing ground clutter interference.

Since the beam scans round, continuous readings are not made in any one direction; instead a brief glimpse is obtained as the beam passes over. The measurements are thus instantaneous readings of intensity in any one direction taken about once every three minutes. This snapshot view inevitably adds (perhaps 10%) to the uncertainty of conversion from spot observations to mean areal values, which arises from the temporal and spatial variability of rainfall, which in turn depends on the type of precipitation.

Accuracy and corrections

The accuracy achieved by the radar system operated on Hameldon Hill for rainfall measurements was evaluated in detail by comparing the hourly values of the radar estimate with hourly raingauge readings (Collier 1986), and it was found that the differences between the two varied considerably. When observations influenced by the bright band were excluded, the average difference between radar and raingauge readings under conditions of frontal rain was 60%, while for convective rain it was 37%. If bright-band episodes were not excluded, the difference could be as high as 100%.

To help correct these large errors, ground-truth from telemetering raingauges can be used to adjust the value of the factor *a* (in the empirical formula above). However, if only a few gauges are used to give an average correction for the whole area, little is gained because of spatial and temporal variations. An alternative would be to use a large number of gauges to sense the spatial variations more widely, but this requires too many gauges.

A less direct, simpler and cheaper method of correction is based on the automatic detection of the type of rainfall, this allowing the most appropriate value of *a* to be used. Detection of rainfall type can be achieved by measuring the variability between raingauge readings at a small number of sites. This can identify the type of rainfall well enough for the purpose, and can be used to choose the best *a* factor. This approach lessens the difference between radar estimates and raingauge measurements from 60% to 45% for frontal rain and from 37% to 21% for convective rainfall.

Corrections can be made for permanent ground obstructions and spurious ground clutter reflections since they do not move and are present in dry weather. Allowance is also made for a reduction in outgoing and returned signals due to their absorption as they pass through heavy rain. At longer ranges, allowance also has to be made for earth curvature and for the fact that the beam may, therefore, be above at least some of the rain. Radar measurements are thus not as precise as a good raingauge network can supply, but where such a network is not available radar is invaluable.

New developments

Although operational weather radar networks are now well established, research continues, both from the instrumental point of view and also as regards the improvement of algorithms. For example, because large raindrops become flattened as they fall they offer a non-circular profile, and if the polarisation of the radiated beam is switched rapidly from the vertical to the horizontal, the ratio of the returned signals from the two polarisations is different, giving what is termed the *differential reflectivity*. This enables drop sizes to be determined, since there is greater reflection when the polarisation is in line with the longer (horizontal) axis of the drops. Small drops are nearly spherical, producing little difference in reflectivity (Hall *et al.* 1980, Hall 1984). This has required both new hardware and software development.

A few large hailstones can give as large a signal as heavy rainfall, although the hail represents much less equivalent water. An experimental *differential phase* method operates by measuring the difference in speed between the vertical and horizontal polarised pulses as they pass through the precipitation. If the precipitation is heavy rain, the horizontally polarised pulses are slowed more than the vertical, because the longer axis of the large drops is horizontal and the wave is slowed down as it passes through more water. (The speed of an electromagnetic wave is slower in water than in air.) This reduction in propagation speed causes a shift of phase of the returned signal, which increases with range (since more water is passed through). For rain, there is a unique relationship between differential reflectivity and differential phase because the drops always have a clearly defined shape, set by size, while for hail the phase shift is much less because the hail is random in shape and alignment (Smyth *et al.* 1999). Thus the presence of hail can be detected and the associated errors avoided.

These more sophisticated techniques could improve the performance of operational radar estimates, but there are problems, not the least being an increase in the complexity of data processing and hardware, and thus of cost. For these reasons this type of radar has not yet been introduced operationally. For those wishing to know more, there is a detailed review by Collier (2002).

And research continues today, for example at the Chilbolton Advanced Meteorological Radar system in Hampshire (operated by the Rutherford Appleton Laboratory at Chilton in Oxfordshire). This field test site was set up specifically to investigate the effects of weather, in particular precipitation, on radio communication, both terrestrial and satellite. The 3 GHz (10 cm) 25-metre antenna (Fig. 10.3) (Kilburn *et al.* 2000) is the largest steerable meteorological radar in the world and has been operating for 30 years. It produces a very narrow beam (0.25 degrees – a quarter the width of operational weather radars) transmitting half-microsecond, 700-kW, pulses at a repetition rate of 610 per second, the returned signals being averaged every quarter of a second and producing a picture of the meteorological conditions up to 50 km distant.

A *cloud radar* is also operated at Chilbolton, in cooperation with Reading University, the higher frequencies of 35 and 94 GHz measuring the smaller cloud droplets, thereby giving an overall picture of the atmosphere at many particle sizes (Hogan and Illingworth 1999). These small antennas can be attached to the 25-metre dish (right of the main dish in Fig. 10.3), operating in line with it at the same time, or they can be operated separately, looking vertically for long periods (Fig. 10.4). In the latter mode, they have been used to evaluate how successfully forecast models predict clouds, comparisons of the radar observations with model predictions suggesting that the models are not yet able to simulate clouds in fine detail, although large-scale features are more precisely modelled.

Figure 10.3 This 25-metre, 3 GHz radar at Chilbolton in the UK is used for research into the effects of weather, in particular precipitation, on radio communication, both terrestrial and satellite. It is the largest steerable meteorological radar in the world, measuring raindrops and clear-air turbulence. To the right are two much smaller antennas operating at 35 and 94 GHz for the measurement of cloud droplets. Photograph courtesy of the Council for the Central Laboratory of the Research Councils (CCLRC).

Figure 10.4 The small radar dishes seen to the right of the main dish in Figure 10.3 can also be operated independently, looking up vertically for long periods to detect cloud amount. They are seen here in the workshops at Chilbolton being serviced.

Disdrometers (see Chapter 8) are also used in some research situations for measuring the actual drops as they fall in real time, providing information on drop-size distribution and helping to improve the algorithms and rainfall intensity estimates.

Nimrod and Gandolf

Another approach to the enhancement of radar rainfall measurements was begun some time ago in the UK under the name of FRONTIERS (forecasting rain optimised using new techniques of interactively enhanced radar and satellite data). Now known as *Nimrod*, a more recent version of this *nowcasting* technique ('Nowcasting' is forecasting from now to three hours ahead) combines IR and Visible Meteosat cloud images (see next chapter), surface weather reports, data from telemetering raingauges, input from numerical models, with the measurements from the 13 radars around the UK. This combination enables judgement to be made (automatically) as to whether a radar echo is from rainfall or from spurious echoes due to ground clutter or 'anaprop' (anomalous propagation, for example due to a hydrolapse), allowing the radar data to be more accurately converted to rainfall intensity than when they are used alone in isolation. Nimrod also goes some way towards compensating for the gradual increase of the height of the beam with increasing range. Estimates from Nimrod are compared with measurements from the raingauge records on a weekly or monthly basis and adjustments made to the radar data to exclude errors introduced by the different sampling characteristics of the radar and the raingauges. Remaining differences in monthly rainfall maps are due to errors in the correction process, radar hardware faults and the inherent shortcomings of the radar technique.

An extension of the Nimrod system, *Gandolf* (generating advanced nowcasts for deployment in operational land-surface flood forecasts), uses satellite images to differentiate between convective and frontal rain clouds, a separate forecast being produced for the convective rain. This is then combined with the forecast of frontal rain using Nimrod techniques, producing a forecast for rain rates and totals for each 15 minutes up to three hours ahead.

This combination of several different methods of measurement is returned to in the next chapter in the correction of remotely sensed measurements from satellites. Such combinations will undoubtedly be an important development in all remote sensing in the future, not only for precipitation but for all the other meteorological variables. This can be seen by looking first at the weather radar images on the UK Met Office web site and then at the Meteosat IR and visible images (see Chapter 11 and Fig. 4.14) on the same web site, the combination of the two clearly being more informative than either alone.

Snowfall measurement by radar

Since snowfall is very difficult to measure by other means, it would be useful if radar could do it better. The same empirical equation is used for snow as for rain, and as in the case of rainfall there is considerable variability in the values of a and b, a lying somewhere between 2100 for wet snow and 540 for dry. However, the same problem exists for radar as for optical raingauges – the size and density, and thus the water equivalent, of snowflakes vary so much that there is a considerably larger margin of uncertainty than for rain.

There is also the additional problem that because it is extremely difficult to measure snowfall by any conventional means on the ground there are no reliable ground-truth measurements against which to compare the radar's snow readings in real time in the operational state. It is possible, however, to use measurements of snow made after it has fallen to assess the radar's performance in retrospect during the developmental phase and so improve the algorithms. Because snowflakes are not spheroid-shaped, polarisation techniques are inappropriate. Snowfall measurement is just a very intractable problem.

Displaying the measurements

Although raw radar images are presented in polar coordinates, they can be converted to a square grid format, a typical grid resolution being from 1 km for ranges up to 75 km to 5 km for ranges from 75 to 200 km, levels of precipitation intensity being displayed as different colours. In addition, rainfall totals can also be calculated and displayed over a period of, say, 15 minutes.

Cost and application

The capital cost of weather radar is high – hundreds of thousands of pounds/dollars per station – and being complex it needs high-level technical expertise to maintain it. Weather radar is thus best suited to situations where measurements are needed from a large area on a permanent basis in real time, such as for a national weather service or a large river authority. Radar is not suited to measuring precipitation over smaller areas or for short durations, although this kind of information may be purchasable by smaller organisations from those operating large radar networks.

A simpler radar instrument

There are now available some simple Doppler radars based on the type of hardware used for speed cameras and traffic-light control. These look upwards and measure the fall speed of the drops and convert this to drop size and thus to intensity. But they are radar in name only and are entirely different to the radars described in the rest of this chapter. It is difficult to get performance figures from manufacturers, but there must be some uncertainty about the conversion to precipitation intensity because the drops (and snowflakes) are of mixed sizes. Their vertical descent speed could also be much affected by wind and turbulence. I mention these instruments for completeness but have never used or tested them; they will undoubtedly find a niche.

References

Collier, C. G. (1985). Accuracy of real-time estimates made using radar. In *Proceedings of the Weather Radar and Flood Warning Symposium*. University of Lancaster.

Collier, C. G. (1986). Accuracy of rainfall estimates by radar, part 1: calibration by telemetering raingauges. *Journal of Hydrology*, **83**, 207–223.

Collier, C. G. (2002). Developments in radar and remote-sensing methods for measuring and forecasting rainfall. *Philosophical Transactions of the Royal Society of London A*, **360**, 1345–1361.

Hall, M. P. M. (1984). A review of the application of multi-parameter radar measurements of precipitation. *Radio Science* **19**, 37–43.

Hall, M. P. M., Cherry, S. M., Goddard, J. W. F. and Kennedy, G. R. (1980). Rain drop sizes and rainfall rate measured by dual-polarization radar. *Nature*, **285**, 195–198.

Hogan, R. F. and Illingworth, A. J. (1999). Analysis of radar and lidar returns from clouds: implications for the proposed Earth Radiation Mission. CLARE'98 Final Report. In ESTEC International Workshop Proceedings WPP-170, ESA/ESTEC. Noordwijk, September 1999, pp.75–82.

Kilburn, C. D. D., Chapman, D., Illingworth, A. J. and Hogan, R. J. (2000). Weather observations from the Chilbolton Advanced Meteorological Radar. *Weather*, **55**, 352–356.

Kurtyka, J. C. (1953). *Precipitation Measurement Study*. Report of investigation 20, State Water Survey, Illinois. Urbana, IL: Department of Registration and Education.

Smyth, T. J., Blackman, T. M. and Illingworth, A. J. (1999). Observations of oblate hail using dual polarization radar and implications for hail detection schemes. *Quarterly Journal of the Royal Meteorological Society*, **125**, 993–1016.

11

Measuring precipitation from satellites

It is not possible to measure precipitation, or indeed any variable, from a satellite directly; what is measured is electromagnetic radiation, nothing else. Everything has to be inferred indirectly from this through algorithms, some simple, some complex, some accurate, some not.

Usually the radiation is natural – either solar radiation in the visible and near infrared bands reflected or scattered without any change of wavelength by the earth's surface and by the atmosphere and its contents. Or the radiation can be in the thermal infrared and microwave bands, radiated by virtue of the surface and atmosphere having been heated by solar radiation, re-emitting the energy at longer wavelengths. These various wavelengths are measure by different types of instrument aboard different types of satellite. Another class of instrument aboard some satellites actively radiates microwaves and detects the scattered signal just as ground-based radar does. This chapter looks at the satellites, at their instruments and at how the measurements can be converted into estimates of precipitation.

Before starting, it is useful to say something about the satellites themselves.

Satellite orbits

There are three types of orbit into which satellites with remote sensing (RS) and communications capabilities are normally placed – geostationary, polar and skewed.

Geostationary orbits

Satellites with an orbital distance of 35 786 km from the earth circle the earth exactly once a day. If placed in such an orbit around the equator, orbiting in the same direction as the earth rotates, the satellite appears stationary in the sky, and is known as *geostationary equatorial*. More precisely they should be termed *geosynchronous* (Fig. 11.1).

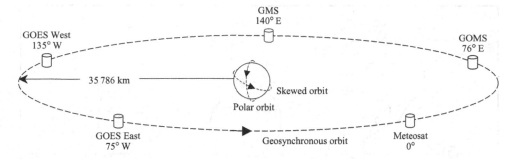

Figure 11.1 Diagram showing the relative positions of the geosynchronous weather satellites and the paths taken by satellites in polar and skewed low-earth orbits.

Polar and skewed orbits

Satellites that orbit the earth over the poles can have any altitude required, provided they are above the atmosphere, because they do not have to keep step with the earth's rotation. Meteorological satellites in polar orbits usually fly at an altitude of about 850 km, giving an orbital period of around 100 minutes: these are known as low-earth orbits (LEOs). The plane of these orbits is usually kept facing the sun, the earth rotating beneath, the orbits being known as sun-synchronous. The satellite passes over a different swath of ground at each orbit, gradually scanning the whole of the globe, the sun always being at the same angle at each pass. While this gives some consistency, in that the sun angle is always the same at each overpass, it does not provide a view of the earth at different times of the day, which is a disadvantage when precipitation and other variables are being measured because of their diurnal variations. LEOs can also be skewed, the satellite orbiting at any desired angle relative to the poles (Fig. 11.1).

Orbits compared

Geostationary orbits present the satellite with the same view of the earth (*footprint*) all the time, the whole disc of the globe being visible (Fig. 11.2). Immediately beneath the satellite, the earth's surface is seen square-on, with a pixel resolution typically, in the visible wavelengths, of 2.5 km × 2.5 km, while in the infrared it is 5 km × 5 km. In all other directions, away from the nadir point, the view is increasingly oblique, to a point where there is no view at all. In particular, the high polar regions can never be seen from geostationary satellites, although telemetry communication via Meteosat is possible as far south as the Antarctic Peninsula. I have done this myself and can confirm that the link is stable and reliable (see Fig. 8.13).

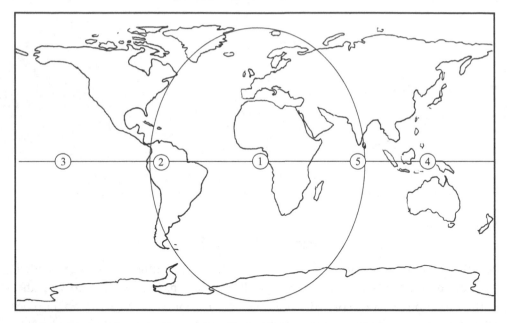

Figure 11.2 The footprint of the European Meteosat satellite centres over (1). (2) and (3) are the centre points of the American GOES E and GOES W satellites. (4) is the location of the Japanese GMS and (5) the Russian GOMs satellites. All the footprints are the same size.

Polar orbits, in contrast, give an ever-changing view of the surface, the satellite sweeping out a swath up and down the earth and over the poles, each part of the earth being viewed more or less vertically downwards and to higher resolution because of the lower altitude. However, repeat images may be many hours, even days, apart.

While polar satellites pass over the poles at every orbit, at any point on the equator a polar satellite is only available about four times a day – even though it crosses the equator the same number of times as it crosses the poles. With an orbital time of 100 minutes there are around 14 passes a day at the poles. Satellites in skewed orbits do not pass over the poles, and as such orbits are not sun-synchronous they are able to collect images at different times of the day at each overpass.

The satellites

The very first satellite to be put into orbit was the Russian Sputnik 1 in 1957, but it did not make measurements. The Americans launched the first weather satellite, TIROS 1 (Television and Infra-Red Observing Satellite), in 1960, and ITOS (Improved TIROS Operational Satellite) was launched in 1970 (it is interesting to reflect that we went to the moon in between). To this point, vidicon television cameras were used to get the images, but in 1972 they were replaced by the first scanning radiometers on NOAA-2. There then followed Landsat 1 in 1972, Skylab in 1973, and Nimbus 7 and Seasat in 1978. These were all in polar orbits. The Russian Meteor weather satellites are another set in polar orbit. EUMETSAT (the organisation managing the European meteorological satellites) also planned to launch the first of three polar-orbiting satellites (MetOp) in July 2006, and these will produce high-definition images (EUMETSAT 2006). (Launch postponed.)

In 1975, the first of the geostationary satellites, GOES-1, was launched, followed in 1977 by the European Meteosat and then by the Japanese and Russian equivalents, GMS and GOMS.

There are also many national satellites, such as India's geostationary Insat and Brazil's SCD-1 satellite. The latter is in equatorial orbit with an inclination of 25°, giving coverage of all of Brazil and of the entire world's tropical and subtropical countries.

Of particular importance to us are the satellites in skewed LEOs designed specifically to measure precipitation – the Tropical Rainfall Measuring Mission satellite and the Global Precipitation Measurement satellite – of which more later.

The electromagnetic spectrum

Figure 11.3 illustrates the electromagnetic spectrum, with the extreme ends (low radio frequencies and gamma rays) omitted because they are not used in RS from satellites as yet. Approximate indications of the bands of radiation that are absorbed and transmitted by the earth's atmosphere are also shown in the figure. These 'windows' are important, since RS of the surface is only possible from space in the bands that are not absorbed by the atmosphere, in particular the visible and the two infrared windows A and B, where transmittance is high.

The bands that are absorbed are also important, however, because they radiate as well as absorb, and this can be used to obtain profiles of temperature and water vapour and other variables down through the depth of the atmosphere from sounders or profilers – of which more later.

The most-used wavebands

The largest concentration of measurements from satellites is in the visible region (0.4 to 0.75 μm), with slightly fewer in the near infrared (0.75 to 1.1 μm). The next most common band is the thermal infrared window marked B in the transmittance graph of Fig. 11.3 (10.5 to 12.5 μm). In the region between 6 and 7 μm water vapour

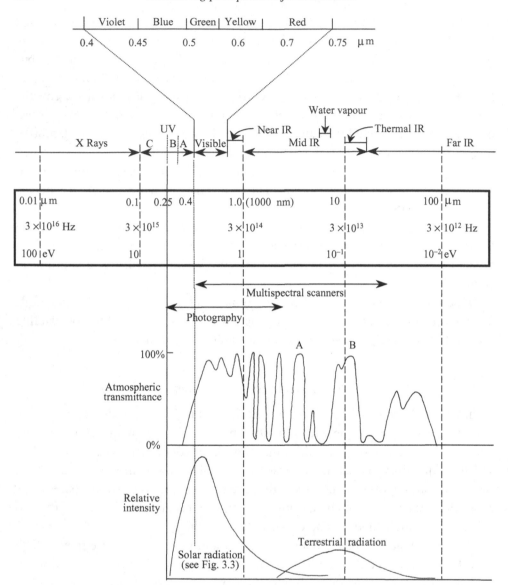

Figure 11.3 The electromagnetic spectrum is shown here from X rays to VHF radio. The long-wavelength end of the spectrum continues further down into extremely low frequencies (see Chapter 6 on lightning) while at the top end it continues into gamma rays, but neither of these extremes is used in satellite remote sensing. Also shown are approximate curves of atmospheric transmittance and the spectrum of the sun and infrared radiation from the ground and atmosphere (see also Fig. 3.3).

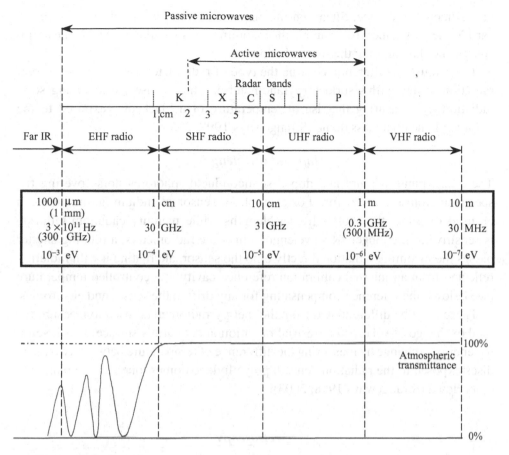

Figure 11.3 (cont.)

radiates strongly, and a few satellites (such as Meteosat) make measurements of humidity by sensing this frequency band.

In the microwave region, passive sensors measure over the range 10 to 85 GHz (see Special Sensor Microwave Imager (SSM/I) and Tropical Rainfall Measuring Mission Microwave Imager (TMI), later).

Sensing the visible and infrared spectrum

Radiometers

Visible wavelengths

Visible-light radiometers measure the radiation within a fixed angle of view, either a wide angle, averaging the whole scene (as in camera exposure meters) or in a highly focused way, using lenses for spot measurements. The light is directed onto a sensor such as a photoresistive cell of cadmium sulphide (sensitivity 0.5–0.9 μm) or a silicon photodiode. Such non-imaging radiometers are used to measure the earth's energy balance, for example the incoming solar irradiance outside the atmosphere and the outgoing thermal radiation.

These must be differentiated from the type of radiometer used to measure solar radiation at the earth's surface (pyranometers), which sense the incoming solar radiation by measuring the rise of temperature of a black disc exposed to the radiation beneath a glass dome (Strangeways 1998, 2003).

Infrared wavelengths

There is a range of pure and doped semiconductor photoresistors covering the spectrum from 0.5 to 20 µm. For example, a sensor of indium antimonide can be used to detect the shorter IR wavelengths while mercury cadmium telluride is sensitive to the longer IR wavelengths. In some radiometers, a rotating shutter passes the incoming radiation directly onto the sensor, alternating it with radiation reflected from an internal calibration reference cavity at a controlled temperature (see below), the reference compensating for any drift in the sensor and electronics.

These must be differentiated from the 'net pyrradiometers' used to measure the total exchange of solar and terrestrial radiation at the earth's surface, which sense the energy exchange by measuring the difference of temperature between two black discs exposed to the radiation beneath polyethylene domes, one facing up and the other down (Strangeways 1998, 2003).

Spectroscopy

By inserting a filter into the radiation path of a visible or infrared radiometer, the beam can be split into discrete spectral bands and the intensity of each measured separately. By using a variable interference filter it is possible to divide up a wide spectrum into many very narrow (5 µm wide) bands.

But these types of simple radiometer only look at one small point at a time, or encompass very wide angles of view, and so are limited in what they can achieve. If they are made to scan a complete scene, they become much more useful.

Scanning radiometers

To obtain a high-resolution image of a wide scene it is necessary for the radiometer to scan the view both vertically and horizontally.

Scanning radiometers in polar orbits

In the case of a geostationary satellite such as Meteosat, both the x- and y-axes have to be scanned, but if the satellite is moving relative to the ground, as in a polar or

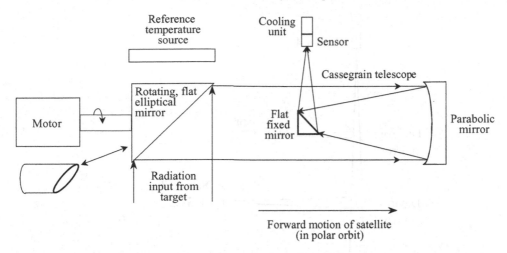

Figure 11.4 A scanning radiometer in a polar orbit uses the forward motion of the satellite to produce the *x*-scan while the rotating mirror generates the *y*-scan. The radiation is reflected into a small Cassegrain telescope which focuses the radiation onto a sensor via a 45-degree mirror. When the rotating mirror points upwards it receives radiation from a reference temperature source – which can be deep space at near absolute zero.

skewed LEO, the forward (along-track) motion of the satellite provides the *x*-scan and only the *y*-axis has to be scanned by the instrument. In this case the frequency of the across-track scan and the width of the scan strip are chosen to ensure that adjacent individual strips neither overlap each other nor have gaps between them.

The most usual way of scanning the scene is to rotate a mirror at 45 degrees to the horizontal at several hundred rpm by means of a motor, the radiation being reflected into the system's optics, typically a small Cassegrain telescope (Fig. 11.4). When the mirror is pointing downwards it scans the earth, when it looks up it reflects radiation from an on-board calibration source. It can also look at deep space, at a few degrees above absolute zero, as an additional reference.

Scanning radiometers in geostationary orbits

Aboard a geostationary satellite such as Meteosat or GOES, where there is no motion of the satellite relative to the ground, radiometers have to scan in both the *x* and *y* directions. To achieve this, the whole satellite rotates on its axis at 100 rpm, the spin creating the *x*-scan, while a mirror steps down by a very small angle at each rotation, giving the *y*-scan. In this way, over a period of 25 minutes, the satellite scans the entire visible globe, producing 2500 scan lines. (To compensate for the rotation of the satellite, its antennas are electronically switched so that continuous radio communication is maintained with the ground station.)

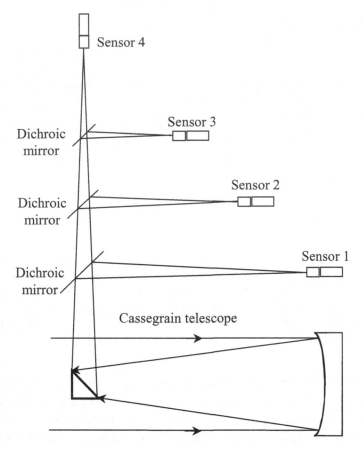

Figure 11.5 A multispectral scanner (MSS) has a number of dichroic mirrors inserted in the path of the beam emerging from the telescope, the radiation being directed to several sensors, each measuring a specific band of radiation.

Multispectral optical scanners

Multispectral scanners (MSSs) represent the next step-up in complexity, combining a scanning radiometer with spectroscopy. Instead of radiation from the Cassegrain telescope being directed to a single radiation sensor, it is passed through several dichroic mirrors or through a mechanically switched filter. The radiation from each is focused onto a separate sensor, sensitive to a particular part of the spectrum, possibly with an additional filter to modify the sensor's range of spectral response (Fig. 11.5).

One of the best known multispectral scanners in a polar orbit is the Advanced Very High Resolution Radiometer (AVHRR) operated by NOAA. The satellite flies at an altitude of 833 km, giving a resolution of 1.1 km at the nadir point, covering a swath of 2600 km width. Five channels are sensed, one in the visible (0.58–0.68 μm) for daytime cloud mapping, one in the near IR (0.725–1.00) for surface water and ice, with three in the thermal IR (3.55–3.93, 10.30–11.30, 11.50–12.50 μm) used for precipitation and sea surface temperature (SST) measurements and for night-time cloud mapping.

Typical multispectral scanners in geostationary orbit are those aboard Meteosat, GOMS, GMS and GOES, the latter being familiar as the Visible and Infrared Spin-Scan Radiometer (VISSR). The first Meteosat second generation (MSG-1) Spinning Enhanced Visible and Infrared Imager (SEVIRI) became operational on Meteosat 8 on 29 January 2004; others will follow, possibly up to four in total. The new satellites will make observations in 12 channels from the visible (one visible channel having a resolution of 1 km – very good from a geostationary orbit – the remainder are 3 km) through the near IR to the far IR at 12.5 μm. The various spectral bands will be used to measure many different variables, including precipitation, fog and depressions.

Passive microwave imagers

Just as in the visible and IR spectrum, microwaves can also be sensed by scanning the scene. In this case, however, the optical methods of mirrors and glass have to be replaced by radio techniques, a small parabolic dish focusing the microwave radiation into one or several feed horns, each sensitive to particular frequencies and polarisations. The horns contain a diode sensor to convert the microwave energy into an electrical signal, which can then be amplified and digitised.

Tropical Rainfall Measuring Mission Microwave Imager

The best-known, and original, passive microwave imager is the Special Sensor Microwave Imager (SSM/I) developed by Boeing, which has been flying continuously in polar orbit since 1987 with around six replacement satellites. It is about 2 metres in height with a dish of 60 cm diameter which rotates about a vertical axis at 31 rpm, sweeping out a swath 1400 km wide at each rotation. The system is multispectral, measuring microwaves emitted at 19.35, 22.235, 37.0 and 85.5 GHz in both vertical and horizontal polarisation, except the 22.235 GHz (oxygen) band, which is vertical only. At these frequencies, with a small dish and flying at an altitude of 860 km, the resolution on the ground immediately below the satellite is 68×43, 60×40, 37×28 and 15×13 km respectively at the four frequencies.

NASA, the National Space Development Agency of Japan (NASDA) and the Japan Aerospace and Exploration Agency (JAXA) later commissioned Boeing to improve the SSM/I, resulting in the Tropical Rainfall Measuring Mission Microwave Imager (TMI). This was launched in 1997 aboard the Tropical

Rainfall Measuring Mission (TRMM) spacecraft built by NASA's Goddard Space Flight Centre, the satellite flying at an altitude of 350 km in a circular, non-sun-synchronous, precessing, low-inclination (35°) skewed orbit with a period of 91.5 minutes, giving about 15 orbits a day. This allows high resolution as well as the ability to see diurnal variations in precipitation (which sun-synchronous orbits do not). In addition to the TMI, the satellite also carries a visible and IR scanning radiometer (as described earlier), sensors for the Clouds and the Earth's Radiant Energy System (CERES), a Lightning Imaging Sensor (see Chapter 6) and, importantly, a precipitation radar (PR) – of which more will be said later under the heading of 'active microwave sensing'.

The TMI antenna is an offset parabola (tilting downwards at about 30 degrees from the vertical) with a diameter of 61 cm and a focal length of 50.8 cm, rotating around a vertical axis directing the radiation into similar horn detectors as are used in the SSM/I. Figure 11.6 shows the area swept out as the satellite moves forward and the dish rotates. This method of scanning is known as 'conical scanning' to differentiate it from the across-track scanning used with the optical radiometers and radar. The conical method has the advantage that the angle at which the ground is viewed is constant, while in the across-track method the angle varies. This is of benefit because emissivity varies with angle.

The TMI measures similar frequencies to the SSM/I but with the addition of 10.7 GHz, all with dual polarisation except the 21.3 GHz channel. The 10.7 GHz channel was added to improve measurement of heavy rainfall. These frequencies are used because they are the best suited to measuring specific atmospheric and oceanographic variables (as will be shown below, in the section dealing with how the measurements are converted to precipitation rates).

Resolution varies from 5 km at 85.5 GHz (sensitive to ice precipitation particles), to 45 km at 10.7 GHz (sensitive to water droplets). This is an improvement on the resolution of the SSM/I, not through better instrumentation but simply because the satellite carrying the instrument flies at the lower altitude of 350 km compared with the 860 km of the SSM/I. However, to extend its life from the expected three years, the satellite was raised to a higher orbit of 402 km in 2001 to reduce atmospheric drag and thus conserve what fuel was left, its resolution being lessened thereby.

How precipitation is measured by the TRMM instruments is discussed later.

Global Precipitation Measurement mission imager

The TRMM was due to be decommissioned during 2005 and a successor, the Global Precipitation Measurement (GPM) mission, was planned to replace it. It would appear, however (at the time of writing in early 2006), that the GPM satellite will not now be launched until 2011 at the earliest, due to revised priorities at

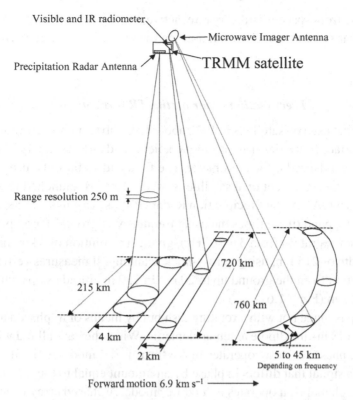

Figure 11.6 The Tropical Rainfall Measuring Mission (TRMM) satellite carries three instruments to measure precipitation – the visible and infrared scanning radiometer, a passive microwave imager and an active microwave precipitation radar. These are described in the text. The figure illustrates the methods of scanning and the dimensions of the areas covered.

NASA. A decision has, therefore, been taken by NASA to keep the TRMM satellite flying for several more years by abandoning the original intent to decommission it with a planned re-entry (which consumes its onboard fuel). This will allow the remaining fuel to be used to keep the TRMM satellite in the correct orbit perhaps until 2010 or 2012; a non-controlled re-entry poses negligible risk to people and property on the ground. GPM will use a new imager named the Conical Scanning Microwave Imager/Sounder (CMIS) with a larger (2 metre) dish that will give improved resolution and measurement accuracy.

Active microwave sensing

All the above instruments measure radiation of natural origin. There are, however, advantages in emitting radiation actively from the platform and measuring the returned scattered signal. Although lidar is becoming available on satellites, active lidar sensing from space is still very rare, active sensing being based on radar transmitters. Lidar systems are, however, used on the ground as sounders (Strangeways 2003).

Precipitation radar on the TRMM satellite

Radar was first used on satellites in 1972 aboard Skylab to measure altitude, notably of the sea surface (not as level as one might believe) and on Seasat in 1978 to measure sea-surface windspeed by scatterometer. The first, and so far only, use of radar for precipitation measurement on a satellite is on the TRMM, launched in 1997.

Designed by JAXA, the Precipitation Radar (PR) uses a single frequency of 13.8 GHz in the K_u band (three times the usual frequency of ground-based precipitation radars, which operate at from 3 to 6 GHz), giving a resolution of 4 km diameter and a swath width of 215 km, using across-track scanning. It measures vertical profiles of precipitation from the ground up to 20 km in 250 m altitude steps when looking vertically down (Fig. 11.6).

Scanning is made not with a rotating dish but by means of a 'phased array'. This consists of 128 in-line dipole antenna elements. When they are all fed with a signal of the same phase the array operates in the 'broadside' mode, while if each dipole is fed with a signal that differs in phase by an amount equal to the electric distance between the elements, it operates in 'endfire' mode. A phased array is thus a linear-array antenna in which the phase of the feed to each element is adjusted so that the angle of maximum radiation (and reception) can be varied, thereby achieving a scanning effect without the need for any mechanical movement. By this means a cross-track scan is completed in 0.6 seconds. The beam width is 0.71 degrees, the pulse length 1.6 μs and the power 500 watts.

However, the PR is limited to measuring rainfall more intense than 0.7 mm per hour, below which it is not detected. In the tropics, and with convective rain, this is not a problem, but in other regions and with other types of rain it is a limitation. How rainfall is measured using PR is discussed later.

Precipitation radar on the GPM satellite

As noted earlier, the TRMM was due to be decommissioned in 2005, to be replaced by its successor, the GPM mission. While under revised current plans GPM will not become operational until at least 2011 it is worth looking at what it will then do. Until GPM is launched, the TRMM satellite will continue to operate, and there could even be a short overlap that will allow readings to be compared.

The GPM satellite will carry a PR working on two frequencies – the present 13.8 (as aboard the TRMM satellite) and an additional 35 GHz, the original lower frequency being unable to measure light rain and snow, while the new higher frequency can, although it cannot measure heavier rainfall. The two frequencies together will thus cover a wider range of precipitation intensity than the current TRMM.

The PR together with the Conical Scanning Microwave Imager/Sounder (CMIS), also known as the Global Precipitation Measurement Microwave Imager (GMI) will give improved measurements of rainfall, rainfall processes and cloud dynamics. The Core spacecraft will fly in a 65-degree inclination orbit at a height of 407 km, giving coverage over most of the globe where liquid precipitation occurs.

To achieve a three-hourly revisit time, a number of satellites will need to be used: NASA estimate that eight will be sufficient. The Core spacecraft will be the central element of the GPM space segment, deploying the passive microwave imager (GMI) and the PR, with other GMIs being operated on NASA's Constellation spacecraft. 'Constellation' is the term used to describe the group of additional spacecraft that will supplement the microwave measurements from the core satellite. The constellation satellites will be launched by other US agencies and their international partners, some already being in operation. These partners include the European Space Agency and joint French and Indian satellites. The PR will not be duplicated on the constellation satellites and will be used mainly to calibrate the passive microwave measurements.

Limitations of satellite measurements

Even with the three-hourly revisit times planned for the GPM satellites, only a rapid occasional snapshot of precipitation will be obtained, and this falls short of what is required to give adequate cover. Furthermore, with the coarse resolution of the instruments, small rain events will be missed. A solution to the low frequency of observation would be to put microwave imagers on geostationary satellites, from which revisit times of 30 minutes would be possible. This would complement the visible and IR sensor readings already made from the satellites. But such instruments are not yet available and, at the distance involved, resolution would be less than in LEO.

A further problem with satellite measurements is that of the discontinuity of records due to platform and sensor changes without overlapping intercomparisons with the replacement instruments. The lifetime of satellites is only about three to five years, after which wear and tear and fuel consumption mean they need to be replaced. During the three years, calibration can drift, as can the satellite orbits and the equator-crossing time. These problems are well illustrated by the current situation, in which the TRMM is ageing in its much-extended mission.

While they are an important step forward, satellite measurements of precipitation are, therefore, still quite limited.

Sounders

Whereas an imager produces a two-dimensional map of the radiation, *sounders* or *profilers* measure a profile vertically beneath the satellite, their main application being to obtain measurements of air temperature and water vapour content at different heights up through the atmosphere from the surface to 40 km altitude by making observations in both the IR and microwave bands at different wavelengths. Data handling and analysis involve a complex mathematical inversion process.

The first Microwave Sounder Unit (MSU) became operational in 1979 on board NOAA's TIROS-N satellite in a polar orbit, and it has been used to estimate precipitation back to its launch date. However, although it provides a longer record than the SSM/I (launched in 1987), the latter has much-improved performance and resolution. The MSU has now evolved into the Advanced Microwave Sounding Unit (AMSU-A), launched in 1998, which has 12 channels in the range 5 to 60 GHz, four around the 183 GHz water vapour line, and four channels at 23.8, 31.4, 50.3 and 89 GHz. Along with its predecessor, the MSU, the AMSU is devoted primarily to temperature soundings, and although the sounder measurements can be used to extract precipitation estimates, scanners such as the SSM/I do the job much more effectively. The principle of data extraction is, however, the same, as summarised later for scanners.

Estimating precipitation from the satellite measurements

We have looked at the instruments aboard the various satellites; let us now look at how their measurements are used to estimate precipitation. Techniques fall into two main categories: an estimation of how much may be falling, using visible and IR wavelengths; and the measurement of how much precipitation is falling, using microwaves, both passive and active. A simpler method has been developed that detects the change in ground condition after rain has fallen, but this is not precise, is limited to arid regions and will not be described further. But first we need to look at how the raw data are pre-processed prior to conversion to precipitation.

Brightness temperature

All satellite images are composed of numerous individual pixels, which when displayed on a screen give the impression to the eye of a recognisable continuous two-dimensional view of the earth. In order to derive numerical information from the image, however, each pixel is treated as a point value (even if it does represent an area as large as 45 km diameter).

Radiometers measure the *radiance*, in specific bands of wavelength, of the emitting surface. In the visible spectrum, the radiance is converted to the *brightness*, or the *reflectance* of the surface, as recognised by the eye. The strength of the received signal in the IR and microwave bands is converted to temperature using the concept of *brightness temperature*, that is the temperature that the target would have to be at to produce the detected level of signal. This needs a brief explanation.

An ideal 'black body' absorbs all the radiation falling on it. Real objects, however, absorb only a fraction of the radiation, expressed as their *absorptivity*. The efficiency of radiation, the *emissivity*, of a body is the same as its absorptivity at the same wavelength. *Radiance* is the power radiated per unit area per steradian per unit wavelength and is related to the temperature of the body radiating it, at a given wavelength, by Planck's radiation law for perfect absorbers or emitters, expressed as:

$$\text{Brightness temperature, } T_B = I(\lambda^2/2k)$$

where I is the radiance, λ is the wavelength and k is Boltzmann's constant.

But surface texture also affects emissivity and there is, therefore, no absolutely precise equivalence between the measured radiation and the temperature of the object emitting it. (In the case of radar, brightness temperature is not involved because it is an active microwave technique, the reflectivity and scattering of the radar pulses being what is measured.)

Brightness temperature then has to be converted to units of the variable of interest, in our case to precipitation. This needs some form of algorithm.

Algorithms

Some variables need not be expressed as a precise quantity, a two-state analysis being sufficient (cloud/no cloud, snow/cloud, sea/land, desert/forest, for example), but other variables require numerical values to be derived, precipitation being one of these. Such variables are less simple to sense remotely because the electromagnetic spectrum reflected from, or emitted by, the earth's surface or by the atmosphere or the atmosphere's contents is rarely directly and simply related to the variable we want to measure.

To overcome this lack of direct equivalence, algorithms have to be developed to convert the raw radiation measurements into the variable and units required. These algorithms can be complex, may not be very precise, or may only be applicable to a particular location, to a particular set of circumstances, or to one season. Most of the work done with regard to measuring precipitation (and indeed all variables that need to be quantified) from satellites has gone into the development of algorithms, and success depends critically on how good these are.

Ground truth

RS far out-performs ground-based measurements in many applications due to its wide areal cover, but where precise quantitative values are needed, as in precipitation, accuracy is limited to the skill of the algorithm, by the physical complexity that needs to be modelled, and by the resolution provided by the instruments. However, it is possible to ameliorate the lack of a direct equivalence between the measured radiation and the variable, as well as the effects of atmospheric interference on the passage of the radiation through it, by combining *in situ* ground-truth raingauge measurements with the satellite observations. Indeed, as was noted in Chapter 10 in the case of radar, it is the combination of ground-truth and remotely sensed data that gives some of the best results.

Ground-truth measurements of precipitation are obtained from raingauges telemetering their data to a central base in real time. Very often the telemetry link can be via the same satellite that also makes the remotely sensed measurements (see for example Fig. 8.13). It should be mentioned again that it is important to ensure that the *in situ* measurements are as good as possible, since the RS measurements are calibrated by them. Thus gauges exposed in a pit at ground level should be used whenever possible and the site must also be properly maintained over the decades.

Cloud identification

Before looking at how satellite measurements are used to estimate precipitation, it is useful to detour slightly to look at how the images can be used to identify clouds from space. Images of clouds are one of the greatest benefits and most obvious features of satellite RS, giving information never before available. EUMETSAT (2000) state that the main mission of Meteosat is to generate 48 cloud images a day for use in short-period forecasting, the visible channel producing images in daytime, the IR during both day and night. The International Satellite Cloud Climatology Programme, part of a WMO research project, produces global maps of monthly averages of cloud amounts.

The methods used to extract cloud information automatically from the thermal IR images are typified by those in which a series of tests is applied to the individual pixels, any colder than a specified brightness temperature being deemed to contain cloud. A second test checks adjacent pixels, a high variance between them suggesting either a mix of clear and cloudy pixels or clouds at different levels. Small variations and low temperature indicates full cloud cover.

By measuring the ratio between the near IR images (0.9 μm) and the visible (0.6 μm) it is possible to establish whether the pixel represents cloud (ratio 1.0), water (0.5) or land (1.5). However, because land surfaces have a wide range of emissivities, these ratios are somewhat uncertain. This test is also used to detect the absence of cloud so that surface variables, such as sea surface temperature, can be measured.

By comparing the brightness temperatures at several different wavelengths, further deductions can be made. For example by comparing the 3.7, 11 and 12 μm IR bands of the AVHRR, thin cirrus cloud can be differentiated from thick low cloud or fog. By the same technique it is also possible to distinguish between snow and cloud, between cumulonimbus, nimbostratus, altocumulus, cumulus and cirrus over the land. There is, however, considerable overlap in many of these cases and thus some uncertainty.

The height of cloud tops can be estimated in several ways, the simplest being to measure the temperature of their tops, from which height can be inferred. While this gives good results for stratus and large cumulus clouds, it is not particularly effective for semi-transparent cirrus or small cumulus.

As noted in the chapter on cloud formation, it is not yet possible to identify cloud type reliably enough from satellite images to allow ground-based observations to be discontinued. But much information can be obtained by practised observers simply by viewing the satellite images.

The presence and extent of snow and ice is one of the few hydrological variables, however, that can be measured from satellites automatically, because, like open water, snow and ice can be recognised on a yes/no basis. Several methods have been developed for differentiating snow from cloud, such as that of Ebert (1987), which uses an algorithm based on albedo thresholds. Dozier (1987) describes a method using the 1.55 to 1.75 μm channel of Landsat, in which snow exhibits a much lower reflectance than cloud. But manual methods such as looking at *changes* in clouds and the ability to identify ground features such as roads and forests are also used.

However, at IR wavelengths, although snow and clouds can be differentiated from each other, the presence of cloud prevents detection of what the snow and ice cover is below the cloud – and snow-covered areas are rather prone to being cloudy. While microwave frequencies offer an all-weather technique for detecting sea ice, based on the high contrast between the unfrozen and frozen sea at these wavelengths, microwave radiometers have only a coarse resolution compared with IR and so can only detect differences over comparatively large areas.

Estimating precipitation from satellite visible and infrared measurements

The first measurements to be made from satellites were in the visible and IR bands, starting in the 1970s, and so were the first to be used to measure precipitation. While they provide the longest run of data and are the simplest to use, the results are not highly accurate. However they fare better over long time-frames compared with instantaneous readings, and this suits them more to climate applications than to real-time rainfall rates for nowcasting. There are three main approaches to using these frequencies for sensing precipitation: *cloud indexing*, *bispectral* methods and *life-history* methods.

Cloud indexing

Many methods of estimating how much rain is likely to be falling have been developed over the last 30 years that involve looking at the clouds themselves from above, rather than at the rain falling from beneath and within them. The techniques use the satellite thermal IR measurements (10.5 to 12.5 μm) of *cloud-top temperature*, which is typically around $-40\,°C$. These are known generically as cloud-indexing methods, and they may be coupled with other 'predictors' such as ground-truth raingauge surface measurements.

The simplest and most widely used was developed by Arkin (1979) for the GARP Atlantic Tropical Experiment (GATE), using the high correlation between radar-estimated precipitation on the ground and the proportion of the cloud area that is colder than $-38\,°C$, measured in the IR. The method is named the GOES *Precipitation Index* (GPI) (Arkin and Meisner 1987) and designates clouds of this temperature as having a constant rain rate of $3\;mm\,h^{-1}$, which is suitable for tropical precipitation, over an area of $2.5° \times 2.5°$ latitude/longitude. It is regularly used for climatological analysis and by the Global Precipitation Climatology Project (see next chapter) for estimating global precipitation for periods from five days (pentads) to one month (World Climate Research Programme 1986, Arkin *et al.* 1994, Huffman *et al.* 1997). The GPI method has been improved recently by calibration with local raingauges and with microwave data. Correlation is only moderate for individual pixels but if several adjoining areas are included, results are improved (Ebert and Manton 1998). The threshold temperature selected to indicate the probability of rain varies with latitude, being 15 degrees lower in mid latitudes (Arkin and Meisner 1987).

These same techniques can also be used with polar-orbiting satellites, which give better spatial resolution, with cover up to the poles, but with much less frequent cover in time. The method is most reliable with tropical cumulonimbus clouds (Fig. 11.7) and less so with stratiform clouds of the higher latitudes. The method has, however,

Figure 11.7 This IR Meteosat image (30 August 2005) shows typical large cumu-lonimbus clouds over equatorial Africa. By measuring the temperature of the tops of this type of cloud using the satellite's IR channel, it is possible to infer precipi-tation intensity beneath them. Copyright EUMETSAT/Met Office 2005. Published by the Met Office.

been in use the longest and so provides the longer run of data, which is invaluable over the oceans.

Bispectral methods

The simple IR-only methods of cloud indexing can be improved by including information from the visible channel. There is a relationship, although not a hard and fast one, between cold, bright clouds and a higher probability of rain. Precipitation is less likely with cold but dull clouds (for example thin cirrus) or with bright but warm clouds (such as stratiform). The RAINSAT technique (Lovejoy and Austin 1979, Bellon *et al.* 1980) uses this technique. Of course the method is not usable at night, whereas the IR methods are.

Life-history methods

Convective clouds evolve relatively quickly (Chapter 4) and their changes can be followed from geostationary satellites with their 30-minute repeat images. Precipi-tation varies with the lifetime of a cumulus or cumulonimbus cloud and being able to determine the part of the life cycle the cloud is in enables the probability and intensity of precipitation to be estimated. The method works reasonably well in the tropics, where convective clouds dominate and are the main source of precipitation

(Fig. 11.7). In higher latitudes, however, when used with stratus clouds, contradictory results can be produced (Levizzani *et al.* 2002). Indeed all of the methods of precipitation detection based on the visible and IR bands tend to be restricted to the tropics. Their advantage, nevertheless, is great, for the methods have allowed many areas to be monitored for the first time.

However, microwave frequencies offer better accuracy, although their resolution is much lower.

Estimating precipitation from passive microwave measurements

Microwaves are better than IR and visible radiation because they can penetrate cloud, coming into direct contact with the liquid and frozen hydrometeors. Following the launch of the SSM/I in 1987 many algorithms were developed to extract precipitation estimates from its observations. These are of two types.

The first is a pragmatic, experimental approach in which microwave brightness temperature is simply related to observed rainfall rates measured by raingauges, the NOAA algorithm (Ferraro *et al.* 1996) being the prime example.

Physically based algorithms are an alternative, the Goddard Profiling (GPROF) algorithm being the best known (Kummerow & Giglio 1994a, 1994b). These attempt to model the physical processes of radiation, absorption and scattering of the microwaves by the ground and by hydrometeors.

Across their full spectrum (from 1 mm to 1 m wavelength, or 300 to 0.3 GHz), microwaves are radiated by all objects above absolute zero, their brightness temperature being calculated using the Planck function (see earlier). Thus the land, the sea, the clouds, the rain and the snow/ice all radiate microwaves while also either absorbing or scattering the radiation from all the other sources. This is summarised in Fig. 11.8, from which it is seen that the strength of the microwave signal received at the satellite dish is an amalgam of radiation from the surface, combined with that from the hydrometeors, less that scattered or absorbed by the precipitation and cloud droplets.

Because the land surface radiates strongly it tends to swamp the radiation from the precipitation, and this is made more complicated by the emissivity of the ground varying widely depending on its nature and condition, making it difficult to apportion the radiation precisely amongst its various sources. Since neither the hydrometeors nor the land has a strong polarisation signature, this path to separation is not available.

Many algorithms have been developed to reduce the interference of the upwelling microwave radiation from the land surface, including empirical and theoretical correction constants that vary from region to region and with meteorological conditions (Weinman and Guetter 1977, Grody 1984, Spencer *et al.* 1989, Kidd 1998). It is also

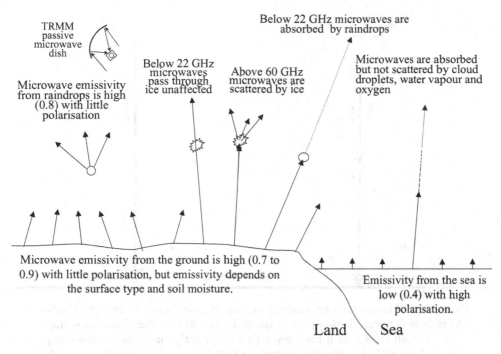

Figure 11.8 Microwaves originating from the ground, from the clouds and from the precipitation all interact with multiple emissions, absorption and scattering. The resultant complex composite microwave signal received at the satellite's scanner is the sum total of all of these. Both the ground and raindrops emit microwaves strongly and it is difficult to differentiate one from the other. Algorithms attempt to take account of these exchanges and separate out the precipitation signal from the background. See text for details.

possible to make corrections by measuring the upwelling radiation from cloud-free areas of adjacent land, although this is not infallible since emissivity is spatially variable and also depends on whether the ground is wet or dry. 'Screens' have been developed that identify different surfaces, for example deserts, forest, sea ice, sea and snow, and make allowances accordingly. Failure to correct for the variations in emission from the land results in it being difficult to be certain if what is detected is rain or uncorrected emission from the ground. Threshold of detection remains a problem.

The sea surface, however, emits at a more consistent and lower intensity as well as being strongly polarised, allowing better separation of the sources. Thus, over the oceans, brightness temperatures at frequencies below about 50 GHz are more or less directly related to rainfall. Much better precipitation data are thus available from over the oceans back as far as the launch of SSM/I in 1987, with data of lesser quality over the oceans since the launch of the MSU in 1979.

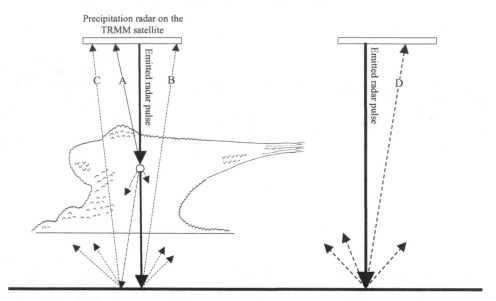

Figure 11.9 Three echoes (or scatters) are produced by active precipitation radar: (A) is from the precipitation back to the receiver, (B) is scatter from the surface, (C) is the rain-echo after it has been reflected off the ground (the 'mirror-image echo'). A measure of the echo strength in rain-free pixels (D) is also made for comparison with (B), to estimate the attenuation of the pulse in its double journey, down and up, through the precipitation. An alternative is to use a second frequency that is not attenuated by the cloud, but this complicates the instrument.

Estimating precipitation from active microwave measurements

Active microwave precipitation radar (PR) is the most important method to have been developed for measuring precipitation from satellites, but at the time of writing only one (that on the TRMM) is in operation.

Although they operate on the same general principles as ground-based radar (Chapter 10), there are some important differences, the most obvious being that a satellite PR is a downward-looking instead of a sideways-looking system. Consequently the echo resulting from the emitted pulses has three components rather than just the one, as shown diagrammatically in Fig. 11.9. As the emitted radar pulse travels downwards, the first echoes are produced from the precipitation itself, followed by an echo from the ground. Finally what is called the 'mirror-image echo' results from the rain echo (scattered downwards) being reflected off the ground and back to the satellite.

The echo from the precipitation is used in the same way as in ground-based weather radar to deduce precipitation (Chapter 10), although since the radar is looking down on the raindrops their profiles will not be visible and so the use of polarised pulses will not be effective. The ground echo is important because it estimates the total attenuation of the pulses as they travel down through the precipitation and back up again through it (Meneghini and Kozu 1990), giving a second estimate of precipitation intensity that is not available in ground-based radar. However, this requires a measure of the echo strength in rain-free pixels, although an alternative is to use another frequency that is not affected by precipitation (the *dual-frequency surface reference technique* (DSRT)). But then the question arises as to whether the second frequency interacts with the surface in exactly the same way as the main frequency. There is also the difficulty that the reflection from the surface will be spatially variable. Reflectance from the sea depends mostly on surface conditions, which are dependent on the wind. Spatial variability of the reflecting properties of the sea will also be less than for the land since it is more homogeneous. Measurements by PR over the oceans are thus likely to be better than those over land.

Comparison of algorithm performance

There have been numerous intercomparisons of satellite precipitation algorithms, notably the NASA WetNet Precipitation Intercomparison Projects (PIP) on the global scale, and on the regional scale the Algorithm Intercomparison Program of the World Climate Research Programme.

Looking at just the microwave imager and PR on the TRMM satellite, it should be expected that there will be a difference in their precipitation estimates because the instruments work on very different principles. In addition, the many algorithms also use different methods and so behave differently. The validation of the various combinations of instrument and algorithm has produced a vast amount of complex literature, but several general points emerge that sum up the situation adequately for our purposes:

(1) No single satellite algorithm performs well in all conditions over all surfaces.
(2) Accuracy varies with time and place.
(3) Microwave methods produce better estimates of instantaneous rainfall than IR.
(4) Causes of variations between microwave algorithms are different over land and sea.
(5) Algorithms that include raingauge ground-truth data give the best results.

Since convective rain is very variable in time and space, one pixel will often contain a mix of convective and stratiform clouds, or just one small convective storm. It is thus difficult, with low-resolution data such as those produced by microwave imagers, sounders and PR, to measure convective rain with precision (Kummerow *et al.* 2001). It is of course also true that the same difficulty applies with raingauges unless the density of the network is high, which in the tropics it is not. When comparing passive and active microwave estimates of precipitation, the general conclusion is that microwave imagers overestimate stratiform and underestimate convective precipitation compared with radar.

In the most extensive comparison of satellite estimates with raingauge ground truth (PIP-3), several algorithms of monthly estimates at 2.5° latitude/longitude grid resolution were compared over 12 months with the Global Precipitation Climatology Centre raingauge dataset (see next chapter). In this programme, 216 land grid cells and 15 tropical Pacific island cells were compared. Over land, using 14 different algorithms, estimates differed by 78%. Over the ocean, 21 algorithms were compared, giving an error of 55%. The best algorithms had errors of 67% over the land and 41% over the sea. Validation and comparison of the many ways of measuring precipitation from satellites is a subject occupying many good minds at the moment. It is an ongoing study and will no doubt continue for a long time. Many of these algorithms are reviewed by Levizzani *et al.* (2002).

Lightning detection

In Chapter 6 the use of the newly developed Lightning Imaging Sensor (LIS) aboard the Tropical Rainfall Measuring Mission (TRMM) satellite was described. There is also a similar device known as an Optical Transient Detector. As the satellite orbits the earth, a map of lightning flashes is built up. This information can be used to verify and enhance the measurements of precipitation made in the visible, infrared and microwave bands. But, as with most satellite measurements, the detection of lightning is not continuous, but simply a brief glimpse as the satellite passes over.

So we are still some way off being able to measure precipitation from satellites to the accuracy required from the global climatological point of view. One great advantage that satellites do have, however, is that they fly over oceans, over the poles and over dysfunctional countries way beneath them. And with satellites no one need get cold or wet or get their shoes muddy. I still prefer the latter, however: it is just more fun. More importantly, raingauges remain the ultimate ground truth for satellite and radar estimates and are the gold standard for all precipitation measurement: hence my efforts to improve their design through R&D field tests (Figs. 8.10 and 8.11) and my wish to see the raingauge network improved, along the lines proposed in the final chapter.

The next chapter looks at the datasets produced from raingauge and satellite measurements and through the merging of the two.

References

Arkin, P. A. (1979). The relationship between fractional coverage of high cloud and rainfall accumulations during GATE over the B-scale array. *Monthly Weather Review*, **106**, 1153–1171.

Arkin, P. A. and Meisner, B. N. (1987). The relationship between large-scale convective rainfall and cold over the western hemisphere during 1982–1984. *Monthly Weather Review*, **115**, 51–74.

Arkin, P. A., Joyce, R. and Janowiak, J. E. (1994). The estimation of global monthly mean rainfall using infrared satellite data. The GOES Precipitation Index (GPI). *Remote Sensing Review*, **11**, 107–124.

Bellon, A., Lovejoy, S. and Austin, G. L. (1980). Combining satellite and radar data for the short-range forecasting of precipitation. *Monthly Weather Review*, **108**, 1554–1556.

Dozier, J. (1987). Remote sensing of snow characteristics in southern Sierra Nevada. In *Proceedings of the Vancouver Symposium on Large Scale Effects of Seasonal Snow Cover*. IAHS Publication 166, pp. 305–314.

Ebert, E. E. (1987). A pattern recognition technique for distinguishing surface and cloud types in polar regions. *Journal of Climate and Applied Meteorology*, **26**, 1412–1427.

Ebert, E. E. and Manton, M. J. (1998). Performance of satellite rainfall estimation algorithms during TOGA COARE. *Journal of the Atmospheric Sciences*, **55**, 1537–1557.

EUMETSAT (2000). *The Meteosat System*. Revision 4. EUM.TD.05. www. eumetsat.int.

EUMETSAT (2006). *The EUMETSAT Polar System*. Brochure EPS.03. www.eumetsat.int.

Ferraro, R. R., Weng, F., Grody, N. C. and Basist, A. (1996). An eight-year (1987–1994) time series of rainfall, clouds, water vapor, snow cover and sea ice derived from SSM/I measurements. *Bulletin of the American Meteorological Society*, **77**, 891–905.

Grody, N. C. (1984). Precipitation monitoring over land from satellites by microwave radiometry. In *Proceedings of the International Geoscience and Remote Sensing Symposium (IGARSS'84)*, Strasbourg, pp. 417–422.

Huffman, G. J., Adler, R. F., Arkin, P. *et al.* (1997). The Global Precipitation Climatology Project (GPCP) combined precipitation data set. *Bulletin of the American Meteorological Society*, **78**, 5–20.

Kidd, C. (1998). On rainfall retrieval using polarization-corrected temperatures. *International Journal of Remote Sensing*, **19**, 981–996.

Kummerow, C. D. and Giglio, L. (1994a). A passive microwave technique for estimating rainfall and vertical structure information from space. Part 1: algorithm description. *Journal of Applied Meteorology*, **33**, 3–18.

Kummerow, C. D. and Giglio, L. (1994b). A passive microwave technique for estimating rainfall and vertical structure information from space. Part 2: applicatios to SSM/I data. *Journal of Applied Meteorology*, **33**, 19–34.

Kummerow, C. D., Hong, Y., Olson, W. S. *et al.* (2001). The evolution of the Goddard Profiling Algorithm (GPROF) for rainfall estimation from passive microwave sensors. *Journal of Applied Meteorology*, **40**, 1801–1820.

Levizzani, V., Amorati, R. and Meneguzzo F. (2002). *A Review of Satellite-Based Rainfall Estimation Methods*. European Commission Project MUSIC Report (EVK1-CT-2000–00058).

Lovejoy, S. and Austin, G. L. (1979). The delineation of rain areas from visible and IR satellite data from GATE and mid-latitudes. *Atmosphere–Ocean*, **17**, 77–92.

Meneghini, R. and Kozu, T. (1990). *Spaceborne Weather Radar*. Boston, MA: Artech House.

Spencer, R. W., Goodman, H. M. and Hood, R. E. (1989). Precipitation retrieval over land and ocean with SSM/I. Part 1: identification and characteristics of the scattering signal. *Journal of Atmospheric and Oceanic Technology*, **6**, 254–273.

Strangeways, I. C. (1998). Back to basics. The met. enclosure, Part 3: radiation. *Weather*, **53** 43–49.

Strangeways, I. C. (2003). *Measuring the Natural Environment*, 2nd edn. Cambridge: Cambridge University Press.

Weinman, J. A. and Guetter, P. J. (1977). Determination of rainfall distributions from microwave radiation measured by the Nimbus-7 ESMR. *Journal of Applied Meteorology*, **16**, 437–442.

World Climate Research Programme (1986). *Global Large-Scale Precipitation Data Sets from the WCRP*. WCP, 111; WMO/TD 94. Geneva: WMO.

Part 4

The global distribution of precipitation

Concern over climate change has generated the need for accurate information on the distribution of precipitation in space and time for climate model evaluation, for the analysis of observed climate change compared with natural variability, and for generating scenarios for climate change impact studies. The information is also important for understanding the hydrological balance on a global scale. The latent heat released upon condensation into clouds is an important energy source in the atmosphere (see Chapter 4), and knowledge of the distribution of precipitation helps in improving weather and climate models.

Raingauge and satellite datasets are the subject of Chapter 12, while Chapter 13 looks at what the datasets tell us about the means and trends in global precipitation. Chapter 14 reviews what the datasets reveal about the variability and extremes of precipitation.

12

Raingauge and satellite datasets

The organisations assembling datasets

Most precipitation data have been collected by national weather services (NWSs) and national hydrological organisations, and by those collecting data for some specific project. But many of these records do not cover long enough periods, few spanning a century without some break or discontinuity. Until concerns about climate change arose recently this was an acceptable state of affairs. Now that we wish to understand and follow the climate changes that have occurred, are occurring and will occur in all climatic variables including precipitation, in the future, it has become necessary to look more closely at all these disparate data and to assemble them into some order covering the entire globe. None of these data were collected for this purpose, but for the simpler and more routine function of weather forecasting on a national scale or for water resources assessment on a local or regional scale. For climatic research, however, it is necessary to assemble all of these unrelated data into one integrated global dataset. Let us first take a look at the organisations involved in attempting this difficult task or in some way contributing to it.

The World Meteorological Organization (WMO)

The WMO is an intergovernmental organisation of 187 member states and territories which originated as the International Meteorological Organization, founded in 1873. Established in 1950, WMO became the specialised agency of the United Nations for weather and climate, operational hydrology and related geophysical sciences. It does not, however, make any measurements itself; instead it acts as a coordinator of the national weather services and related organisations, setting standards, preparing recommendations and producing detailed practical handbooks. It also organises conferences, such as those of its Commission for Instruments and

231

Methods of Observation (CIMO), and promotes, funds and organises instrument intercomparisons.

The Global Climate Observing System (GCOS)

Located at the WMO headquarters in Geneva, GCOS was established in 1992 to ensure that observations and information needed for climate-related issues are obtained. It is co-sponsored by the WMO, the Intergovernmental Oceanographic Commission (IOC) of UNESCO, the United Nations Environment Programme (UNEP) and the International Council of Scientific Unions (ICSU). GCOS is (in its own words) a long-term operational system providing the observations required for monitoring the climate system, for detecting and attributing climate change, for assessing the impacts of climate variability and change, and for supporting research toward improved understanding, modelling and prediction of the climate system. It addresses the total climate system including physical, chemical and biological properties, and atmospheric, oceanic, hydrologic, cryospheric and terrestrial processes.

Like WMO, the GCOS does not make any observations itself, nor does it generate data products, but it 'stimulates, encourages and coordinates' the taking of observations by national and international organisations in pursuance of their own requirements. It provides a framework for integrating observational systems of participating countries and organisations into a comprehensive system focused on the requirements for climate issues, building on existing observing systems such as the Global Ocean Observing System (GOOS), the Global Terrestrial Observing System (GTOS), the Global Observing System (GOS) and the Global Atmospheric Watch (GAW) of the WMO.

The Global Precipitation Climatology Centre (GPCC)

The GPCC is operated by Germany's national meteorological service, the Deutscher Wetterdienst (DWD), and is the German contribution to the World Climate Research Programme (WCRP) and to the GCOS. Their task is to provide gridded datasets of monthly precipitation totals based on observational data. However, they have done this only for the time since satellite data became available; the dataset of combined gauge and satellite data is described below.

The Global Historical Climatology Network (GHCN)

Operated by the National Oceanic and Atmospheric Administration (NOAA), the National Climatic Data Center (NCDC) compiles this dataset (see below) jointly with the Carbon Dioxide Information Analysis Center at Oak Ridge National Laboratory, Tennessee, which is also operated by the NOAA. The NCDC holds the world's largest archive of climate data and is the most useful from our point of view.

Global Precipitation Climatology Project (GPCP)

The World Climate Research Programme (WCRP) was established in 1980 jointly by the International Council for Science (ICSU) and the WMO, joined in 1993 by the Intergovernmental Oceanographic Commission (IOC) of UNESCO. In 1986 the WCRP set up the Global Energy and Water Cycle Experiment (GEWEX), of which the GPCP is a part. Its goals are to develop a more complete understanding of the spatial and temporal patterns of global precipitation.

The Hadley Centre

Part of the UK Met Office, the Hadley Centre in Exeter is one of the world's foremost groups working on the analysis of climate data and the assembly of datasets, its aims being to develop climate models that simulate variability and change over the past 100 years and predict it through the coming century.

Climatic Research Unit (CRU)

The Climatic Research Unit is part of the School of Environmental Sciences at the University of East Anglia, and is another of the world's leading institutions concerned with the study of natural and anthropogenic climate change. The Unit has developed a number of the datasets widely used in climate research, including the global temperature record and a monthly precipitation dataset for global land areas from 1900 to 1998, which is available from their website (www.cru.uea.ac.uk).

We can move on now to see what datasets some of these organisations have produced.

Datasets from raingauges

The number of gauges in operation

Raingauges have been in operation routinely, although initially not in great numbers, since the eighteenth century in the UK and globally since the start of the twentieth century, and thus provide the best run of data available to allow an estimate of global precipitation patterns to be made; indeed it is the only dataset of any length.

Despite all their shortcomings, it is data from these gauges that have been used to estimate annual global means, and to establish how precipitation varies around the globe and how it has changed both in the short and longer term.

In addition to the problems with the raingauges themselves (Chapter 8), there is the added limitation that very few precipitation measurements are made over the oceans due to the difficulty of deploying gauges on ships and buoys, the majority of ocean measurements coming from islands. On land the limitation has been because manual gauges need an operator and so measurements have been confined, until very recently, to inhabited regions of the earth. Nor are countries engaged in civil war or other dysfunctional behaviour likely to have good, or perhaps any, data. Even where data are collected, there is a great variability in their quality from country to country and also within countries, depending on the operator.

Thus only part of the land surface is well covered by gauges, and as land accounts for only 29% of the globe's surface anyway, it is probable that less than 25% of the globe is adequately covered. We are not as well endowed with good precipitation data as might be thought.

Hulme (1995) and Groisman & Legates (1995) estimated that between 200 000 and 250 000 recording gauges had been installed over the previous few decades. In the UK alone about 10 000 gauges have been used by the Met Office to obtain estimates for the period from 1960 to 1990, although not all of the 10 000 were in continuous operation. Nor are data from all of the 250 000 gauges held in one archive and the majority did not, in any case, operate long enough to establish adequate means, which generally requires a 30-year run of data (World Meteorological Organization 1996).

Although the number of gauges now installed is high, the proportion of them producing useful data is small, and growing smaller. In 1900 there were around 5500 usable stations in the world, their number growing steadily to a maximum of about 16 500 in 1966, falling thereafter to 8000 in 1996 and to only 2500 by year 2000 (New *et al.* 2001). The latest number (January 2005) quoted by the GCOS (see above) is just under one thousand. This fall came about through increasing operational and staffing costs, restrictions on the release of the data by national agencies, because the projects that the gauges had been installed for in the first place had ended, because of lack of drive or through political unrest, disease and poverty.

By 'usable gauges' is meant gauges that have been in continuous operation for 30 years at least and that meet basic operational standards. It is not that there are fewer gauges in operation today – there are no doubt many more – it is just that most do not meet the requirements for obtaining long-term means, and the number that do is falling. Numbers may increase partly again, with time, as data work their way through via WMO decadal publications, but the general demise of the gauges is real and will reduce our ability to follow changes in precipitation through the new century. While satellite instrumentation is developing apace (see Chapter 11), satellite measurements alone are not the answer, for the reasons explained in the previous chapter, in particular because of their low spatial and temporal cover.

This reduction in ground-based measurements is wide-ranging, including all climatic data, not just precipitation. The Intergovernmental Panel on Climate Change (IPCC 2001) warns that 'A serious concern is the decline of observational networks.' The report then lists eight broad areas where further work is needed 'to advance our understanding of the climate', the top item in the list being 'to reverse the decline of observational networks in many parts of the world'. It goes on to say that 'Unless networks are significantly improved, it may be difficult or impossible to detect climate change over large parts of the globe.' The IPCC does not make any observations itself, relying on the various national networks to make measurements; it can do nothing but urge, often to little avail – hence its frustration.

Despite all of these shortcomings, it is data from these raingauges that must be used to try to establish how precipitation changed over the twentieth century and how it will do so in the future. A major task in achieving this has been the gathering together of usable raingauge data from the mass of gauges, with all of their discontinuities and errors, which have been operated by diverse organisations since 1900. Having created a dataset it is then necessary to convert the raw data into gridded areal means, and after that to try to establish what the gridded means tell us about global precipitation.

The Global Precipitation Climatology Centre dataset

Probably the most wide-ranging dataset *after* 1985 is that of the GPCC (Rudolf *et al.* 1999), which collects daily and monthly data from the Global Telecommunication System (GTS) – a complex worldwide communication system for the exchange of meteorological data (see the WMO web site or Strangeways 2003). In addition, the GPCC also collects meteorological and hydrological data from over 140 different organisations around the world. What is most important about this dataset is that it includes precipitation data from satellites merged with the raingauge data.

The maximum number of raingauges contributing to the dataset was about 38 000 but these declined (for the reasons explained above) to a core of 7000 GTS stations by 1999 and the number of contributing stations is now only about 1000. The raw raingauge data are not always available because agreement with the organisations supplying the data forbids their disclosure (for political reasons – countries can be sensitive about letting adjoining countries know too much about their water resources). However, when gridded datasets (see below under 'From point

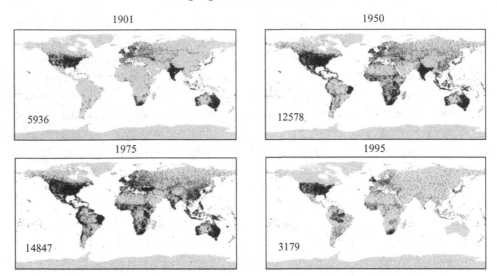

Figure 12.1 The distribution of raingauges used in the GHCN dataset is illustrated by these maps, showing how numbers have changed with time and how some areas are well endowed with gauges while others, especially the oceans, polar regions and the equatorial belt, are poorly represented. From New *et al.* (2001). Copyright Royal Meteorological Society. Reproduced with the permission of John Wiley & Sons on behalf of the RMetS.

measurements to areal estimates') have been derived from the raw station data, they are then available in the public domain. However, the dataset only spans the last few decades and so cannot be used to determine long-term trends or variability.

The Global Historical Climatology Network (GHCN)

An important dataset suitable for establishing long-term trends is the GHCN compiled by the National Climate Data Center (see above), representing a major effort of cooperation between many organisations. The dataset is composed of monthly precipitation measurements over all of the twentieth century, drawing on 31 000 stations with varying time cover, some covering 100 years, others only a few. In 1900 the number of stations was about 5500, increasing to a maximum of around 16 500 in 1966, thereafter falling to about 8000 by 1996 and to 2500 by 2000 (Fig. 12.1). It is this dataset that was used by New *et al.* (2001), and their findings are discussed in the next two chapters.

Errors in the raw raingauge measurements

These matters have already been dealt with in detail in Chapters 8 and 9, but now need to be elaborated on further. Because the datasets are made up of measurements from a wide-ranging and diverse group of gauges, all of different design, exposed differently and maintained and operated to different levels of sophistication, care and technical understanding, the quality of their data vary greatly from one set to another.

Since no site will have remained undisturbed for 100 years, the exposure of the gauges will also have changed over time. The gauges will probably have been changed from time to time, more recently from manual to automatic, all having different outsplash, evaporative and aerodynamic characteristics and errors. It is no easy matter to correct for these and the detailed metadata required to do so are not usually available. New *et al.* (2001) say 'Gradual changes which produce slow changes in the precipitation record are difficult to detect and are generally ignored.' This is an unsatisfactory situation and something that needs attention in the future. (The same applies to changes in temperature measurements, and indeed to all climate variables.)

Adjustments for errors where precipitation falls as snow are extremely difficult due to the aerodynamic effects discussed in Chapter 9. Errors will be large. Where precipitation can fall as either snow or rain, the ratio depending on how warm or cold the winter is, or if there is a gradual transition from snow to rain over decades (due to warming), large errors will result (Legates and Willmott 1990, Førland 1994, Førland & Hanssen-Bauer 2000). There is, therefore, uncertainty as to whether precipitation is increasing in some high latitudes or whether more of it is simply falling as rain (which is caught by raingauges more efficiently than snow).

We must accept, therefore, that the datasets compiled from this wide assortment of gauges from the last 100 to 150 years will be far from precise and will also have changed in precision as time passed. It is not possible, therefore, to obtain as good an estimate of global precipitation means as it has been for temperature, and even the temperature data required a great deal of 'homogenising' to correct for (detectable) errors.

From point measurements to areal estimates

On the local scale

The purpose of measuring rain (and indeed most climate variables) is usually to estimate areal averages rather than to obtain a single-point observation. Over anything other than a very small area, such as one field, several gauges are usually necessary to obtain a reasonable average. While some specialised investigations may use a high density of gauges, even the best meteorological network rarely has more than one gauge for every 25 km^2 (25 $\times$ 10^6 m^2). With a funnel diameter of 12.7 cm (5 inches) the collection area is 126.7 cm^2 (or 0.01267 m^2). This means that the area from which rain is collected is around 1 in 2 $\times$ 10^9 of the area over which rainfall is to be estimated. In most parts of the world the raingauge density is much smaller than this and over the seas it is almost non-existent.

Some have argued with me that in a situation like this, and because precipitation can also be spatially very variable, it is not worth measuring rain very carefully. If raingauge errors were entirely random it may not be quite so vital to ensure high precision since an average of a number of gauge readings over an area would tend to cancel out the errors and thereby reduce the overall error. But in reality the errors all lead to a lower reading than should be obtained, the errors being systematically negative – evaporative loss, outsplash and wind effects; all cause the gauge to catch less than it should. An average of several gauges simply maintains this negative bias.

It is true that if a small convective storm passes directly over a gauge it will catch more than one perhaps a mile away that misses most of it. For one such event, it might look that high raingauge accuracy is, therefore, not important. But over a period of many such storms, random local differences would tend to cancel out and the accuracy of annual and decadal totals of hundreds of such local convective rain events would lead to a precise representation of the sum of the events. But with present raingauges the annual mean is always low, and this does matter.

If we consider single-point measurements, a gauge may catch different amounts to one a few miles away when averaged over a year, and this might be a real difference, perhaps due to topography, or it might be due to different errors of the two gauges. These errors may be due to the gauges being of different designs and so with different errors or it might be that they are the same type of raingauge, but the exposures are different – one in a windier place than the other, for example. We will not, therefore, know if the differences are real or introduced by the gauges unless the gauges are precise. If we are interested in global trends over years, it is equally important to know whether the differences year to year are real or due to raingauge errors (an increasing wind would reduce catch, for example, and vice versa).

Having done everything practicable to get good point measurements, how can these be converted into areal estimates? Imagine an area completely covered in square collecting funnels. The area's average rainfall for any period would be the average reading of all the individual gauges. But in practice most of these gauges will be missing. So how is it possible to estimate the average from just a few gauges?

It would be possible to take the average reading of them all, but there might be several gauges clustered together in one region, or they may be in places that are not typical of the whole area. A technique often used is to work in percentages rather than actual millimetres of rain, each raingauge reading being converted into a percentage of the gauge's average annual catch, determined over a period. This percentage is then converted back to rainfall amounts using the catchment annual average. While this produces more uniform results, the method confers on each raingauge the same importance. One approach to this problem is to give each gauge a weighting that takes account of the relative importance of its contribution to the average (Jones 1983). A widely used technique for doing this is the polygon method (Thiessen 1911), in which the area is divided into polygons such that the polygon around each gauge is that part of the area nearer to that gauge than to any other. The weighting for each gauge is then the ratio of the area of its polygon to the total area. But the gauge surrounded by the largest polygon may not be at a representative site.

Another alternative is to derive the gauge's weighting by dissecting the area into triangles. Briefly the procedure is to construct a mesh over a map of the area and for each mesh point to search for a triangle of raingauges that surround it, but limited to a certain distance. If no such triangle can be found the nearest three gauges are selected. How the mesh is constructed, how the weights are determined and the maximum distance selected, is complex (Jones 1983).

Where a network of raingauges is to be installed from scratch, the opportunity exists to choose the sites. How can this best be done? Gauges could be placed on a regular grid, but this might mean that some are in very unsuitable places such as on a hilltop. In practice it is more realistic to site gauges where there is easy access, or where an observer lives. But a choice made in this way will probably be biased, the remoter, higher and least accessible locations being omitted. The area could first be saturated with gauges, to get somewhere near the ideal of covering the whole area with gauges, irrespective of the difficulty of getting to them. Then, when there has been a chance to study their relative catches, the number of gauges could be reduced, keeping a watch on how this reduction affects the accuracy of the mean areal rainfall. Sometimes gauges turn out to be key ones, being very representative of a large area.

Another technique is to identify topographic domains (Rodda 1962), in which factors such as the slope, aspect and elevation extend over a clearly definable distance, a gauge being placed at random within the domains. When this is done the gauges are installed at ground level in pits with their orifices parallel to the ground (rather than the normal horizontal).

But there is no ideal solution, and the best method for any one situation needs to be selected with an informed knowledge of the options. A combination of two or more of the above methods may be appropriate.

Averaging on the large to global scale over land

The foregoing section is concerned with obtaining a mean precipitation for a relatively small area such as a catchment for water supplies. If the aim is to produce global estimates of precipitation a similar approach can be followed, but for a regular grid rather than an irregular catchment. In the datasets we will look at, this is most usually done by calculating monthly precipitation means for grids of 1° × 1° or 2.5° × 2.5° latitude and longitude, several methods having been developed. There is the obvious problem that as one moves away from the equator, the cells get smaller because the longitude lines converge towards the poles. (Perhaps it would be better to divide the surface of the globe into five- and six-sided polygons, as in the construction of a football.)

Where data are required for analysis of twentieth-century global trends and variability, means are only calculated in cells containing gauges (Hulme 1994) or where there is at least one gauge within an (empirically defined) distance from the grid cell centre (Dai *et al.* 1997). Hulme (1994) uses the Thiessen (1911) method of averaging, referred to earlier. The degree of influence of gauges outside of a cell is weighted according to their distance from the cell-centre by an inverse-distance scheme. The number of cells at which an estimate can be made varies with space and time because of the variability of raingauge positioning and their constantly changing availability. Because of the uneven distribution of gauges, there are also many cells without any mean value whatsoever.

Where the data are needed as input to grid-based simulation models, it is necessary to produce a continuous grid in both space and time over the complete land surface, even though raw data will not exist in many of the cells. In regions lacking data, use has to be made of more remote gauges and this increases the error considerably. Three methods have been developed to achieve this. The GPCP gives monthly precipitation means in grid cells of 1° × 1° or 2.5° × 2.5° latitude and longitude from 1986 to the present (Rudolf *et al.* 1994). A similar dataset has also been produced by Xie and Arkin (1996a) on a 2.5° grid using earlier data extending the series back to 1971, while another is being produced that extends back to 1950. Thirdly, the Climatic Research Unit (CRU) of the University of East Anglia (UEA) has produced a dataset on a 0.5° grid covering 1901–1999, updated annually (New *et al.* 2000). All three datasets are similar, differing only in the number of stations used and the weighting functions applied, the details of which need not concern us here. No entirely satisfactory method has been developed yet, however, to calculate gridded means.

Dealing with a sparse or changing network of raingauges

An important difference between gridding methods is whether the raw basic rainfall depth data are averaged (that is simply a mean of the depth of rainfall in millimetres from all the gauges), or whether the data are first 'transformed'. In the case of the GPCP and Xie datasets, there is no transformation before interpolation, and this can give problems in data-sparse regions which have large variations in rainfall across the area, for example in mountainous regions. The effect of increased elevation is usually to increase the rainfall, but it is not easy to quantify this effect when interpolating the readings over an area in which the height of the ground may vary considerably. This is made more difficult because raingauges in mountainous areas are often in valleys where people live, rather than scattered randomly and at greater heights. A good example of this is Papua New Guinea, where the three gauges representing the country (as listed in the GCOS network) are all below about 45 metres on the coast while the highlands rise to 4500 metres and experience very heavy precipitation including snow.

Working directly in millimetres can also be a problem if the raingauge network changes significantly over time, as it is now doing increasingly. Working in anomalies helps reduce this problem. Where the data are transformed before gridding (Dai *et al.* 1997, New *et al.* 2000), the raw data are first converted to anomalies, relative to a standard normal period, before interpolation. The anomalies may be expressed either in millimetres (Dai) or as percentages (New). These are then converted back to precipitation depth. This is similar to the method described earlier for catchments where percentages are used in place of millimetres, the advantage being that the result is referred back to the long-term past, and to more gauges, thereby having a smoothing effect (Katz and Glantz 1986).

Dai *et al.* (1997) have shown that if the individual station data are first converted into local anomalies, the effect of changing raingauge networks is minimised. They demonstrated this by analysing the global and hemispheric land precipitation using the same raingauge network that was available in 1900, 1930 and at its maximum around 1965. A similar test was also carried out using data from the Sahel region over the period 1920–2004, by comparing results derived from a fixed subset of gauges with those using all available gauges (Dai *et al.* 2004).

These various methods of gridding will naturally lead to different estimates of monthly precipitation. While there is some overlap where the same gauges are used in each method, where there are few gauges the differences increase because of the different ways of selecting which gauges to include in the cells and what weighting to give them, these differences being more prominent where there are only a few gauges. Xie *et al.* (1996), for example, found that when their method was compared with the GPCP gridding technique, the mean absolute differences were about 15% when the 2.5° × 2.5° grids contained five or more gauges, but this increased to between 40 and 90% when there was only one gauge per box. It was also found that when precipitation depth (as raw raingauge readings in millimetres) was interpolated, the means were biased towards overestimating large rainfalls and underestimating small ones, but it is not yet known (at the time of writing) if this is also true when interpolation is via anomalies.

Clearly the conversion from point raingauge measurements to estimated gridded monthly means is not simple and will inevitably be imprecise. Estimating global rainfall is thus not an exact art, but it is improving.

Areal averages over the oceans

A problem already noted in Chapter 8 is the lack of data from over the oceans, in part because of the difficulty of operating raingauges on ships or buoys. In consequence no comprehensive long-term ocean precipitation dataset is available comparable to those for land areas, although some attempts have been made to produce such estimates (Jaeger 1976, Legates and Willmott 1990).

However, a small number of raingauges have been operated on oceanic islands since the 1920s and estimates have been made using 130 cells from island stations across the Pacific, 23 cells from the Indian Ocean and 17 from the Atlantic, between 30° north and 30° south (New *et al.* 2000). An analysis based on this small dataset is discussed in the next chapter.

Rainfall measurements are also made on around 27 moored buoys in the tropical Pacific in connection with the Tropical Atmosphere Ocean/Triangle Trans-Ocean Buoy Network (TAO/TRITON), by NOAA supported by Japan and France, with about 12 further gauges under the Pilot Research Moored Array in the Tropical Atlantic (PIRATA). These measurements are made for investigations into El Niño (see Chapter 14), but buoys cannot be the general solution to precipitation measurement over all of the oceans.

With the coming of satellite measurements, however, the situation regarding the oceans changed and datasets from satellites are now becoming available. However, it is difficult to assess, as yet, the reliability of such satellite observations over the oceans, in part because of the lack of raingauge ground truth to confirm the satellite estimates.

Combined datasets

By combining visible, IR and microwave measurements from satellites with raingauge observations, improved estimates of rainfall can be obtained. To meet the needs of the climate community for accurate estimates of global precipitation, the Global Precipitation Climatology Project was set up to perform this function (Huffman *et al.* 1997). The raingauge dataset used is that produced by the Global Precipitation Climatology Centre, which merged data from around 6000 raingauges and from IR and passive microwave imagers to provide monthly rainfall on a 2.5° grid from 1987 to the present.

Another project that has merged the same data but using more complex statistical techniques is from the NOAA/National Weather Service Climate Prediction Center, known as the Climate Prediction Center Merged Analysis of Precipitation (CMAP) (Xie and Arkin 1996a, 1996b). The period covered is from 1979 to the present.

The merging of data from all the available raingauge and satellite sources in various combinations relies on diverse statistical techniques that are evolving, such as those described by Doherty *et al.* (1999), Huffman *et al.* (2000) and Todd *et al.* (2001). These activities will continue, and improved estimates of global precipitation will emerge.

Some recent work

At the Hadley Centre, long-term climate averages for 13 variables in the UK for the periods 1961–1990 and 1971–2000 have recently been developed, the variables including precipitation totals, rainfall intensity and the number of days of rain with amounts greater than or equal to 0.1, 0.2 and 10 mm of rain, as well as the number of days with hail (Perry and Hollis 2005a). These two authors have also generated monthly and annual gridded (5 × 5 km) datasets for 36 climatic variables, including precipitation, over the UK from 1961 to 2000, the intention being to extend this back to 1914 (Perry and Hollis 2005b). Similar studies are also being done in the USA, for example by Kunkel *et al.* (2003 – see Chapter 14).

Again in the UK, the coordination of national observations of environmental variables is being pursued by the Environment Research Funders' Forum (ERFF), set up in 2002 to reduce overlaps and to fill in gaps in the datasets from all of the UK's major public-sector sponsors of environmental science. Currently all of these data are dispersed amongst the various government agencies in different formats. This is a job of managing existing data; it does not involve the collection of any new data or the analysis of any data. Analysis and interpretation will be the job of researchers who will have access to the new datasets. Details of ERFF can be found at www.erff.org.uk.

In February 2005 the US Environmental Protection Agency set up the Global Earth Observation System of Systems (GEOSS), a cooperative project which currently involves 61 countries, its purpose being to assemble all of the available data under one umbrella. This will be done in cooperation with GCOS, GPCC and the GHCN but it is too early, at the time of writing, to say how this new initiative will develop.

Yang *et al.* (2005) are concerned with correcting, as much as is practicable, the errors in high northern latitude precipitation data due to problems with measuring snowfall. Sparse networks, discontinuities in the records and the under-catch of snow due to wind effects all introduce a large amount of uncertainty.

But let us now move on to the next chapter and see what can be deduced about the worldwide distribution of precipitation using these various datasets.

References

Dai, A., Fung, I. and Del Genio, A. (1997). Surface observed global land precipitation variations during 1900–88. *Journal of Climate*, **10**, 2943–2962.

Dai, A. G., Lamb, P. J., Trenberth, K. E., Hulme, M., Jones, P. D. and Xie, P. P. (2004). The recent Sahel drought is real. *International Journal of Climatology*, **24**, 1323–1331.

Doherty, R. M., Hulme, M. and Jones C. G. (1999). A gridded reconstruction of land and ocean precipitation for the extended tropics from 1974 to 1994. *International Journal of Climatology*, **19**, 119–142.

Førland, E. J. (1994). Trends and Problems in Norwegian Snow Records. In *Climate Variations in Europe*. Proceedings of Workshop on Climate Variations, Majvik, Finland, pp. 205–215.

Førland, E. J. and Hanssen-Bauer, I. (2000). Increased precipitation in the Norwegian arctic: true or false? *Climate Change*, **46**, 485–509.

Groisman, P. Ya. and Legates, D. R. (1995). Documenting and detecting long-term precipitation trends: where we are and what should be done. *Climate Change*, **32**, 601–622.

Huffman, G. J., Adler, R. F., Arkin, P. *et al.* (1997). The Global Precipitation Climatology Project (GPCP) Combined Precipitation Dataset. *Bulletin of the American Meteorological Society*, **78**, 5–20.

Huffman, G. J., Adler, R. F., Morrissey, M. M. *et al.* (2000). Global precipitation at 1 degree daily resolution from multi-satellite observations. *Journal of Hydrometeorology*, **2**, 36–50.

Hulme, M. (1994). Validation of large-scale precipitation fields in general circulation models. In Desbois, M. and Desalmand, F. (Eds.) *Global Precipitation and Climate Change* (ed. M. Desbois and F. Desalmand). Berlin: Springer, pp. 387–406.

Hulme, M. (1995). Estimating global changes in precipitation. *Weather*, **50**, 34–42.

IPCC (Intergovernmental Panel on Climate Change) (2001). *Climate Change 2001: the Scientific Basis* Chapter 14, Advancing our Understanding. www.ipcc.ch.

Jaeger, L. (1976). *Monatskarten des Niederschlags für die ganze Erde*. Berichte des Deutschen Wetterdienstes, 139. Offenbach: Deutscher Wetterdienst.

Jones, S. B. (1983). *The Estimation of Catchments' Average Point Rainfalls*. Institute of Hydrology Report 87. Wallingford: IH.

Katz, R. W. and Glantz, M. H. (1986). Anatomy of a rainfall index. *Monthly Weather Review*, **114**, 764–771.

Legates, D. R. and Willmott, C. J. (1990). Mean seasonal and spatial variability in gauge-corrected global precipitation. *International Journal of Climatology*, **10**, 111–128.

New, M., Hulme, M. and Jones, P. D. (2000). Representing twentieth century space-time climate variability. Part II: development of 1901–1996 monthly grids of terrestrial surface climate. *Journal of Climate*, **13**, 2217–2238.

New, M., Todd, M., Hulme, M. and Jones, P. (2001). Precipitation measurements and trends in the twentieth century. *International Journal of Climatology*, **21**, 1899–1922.

Perry, M. and Hollis, D. (2005a). The development of a new set of long-term climate averages for the UK. *International Journal of Climatology*, **25**, 1023–1039.

Perry, M and Hollis, D. (2005b). The generation of monthly gridded datasets for a range of climate variables over the UK. *International Journal of Climatology*, **25**, 1041–1054.

Rodda, J. C. (1962). An objective method for the assessment of areal rainfall amounts. *Weather*, **17**, 54–59.

Rudolf, B., Hauschild, H., Rueth, W. and Schneider, U. (1994). Terrestrial precipitation analysis: operational method and required density of point measurements. In *Global Precipitation and Climate Change* (ed. M. Desbois and F. Desalmand). Berlin: Springer, pp. 173–186.

Rudolf, B., Gruber, A., Adler, R., Huffman, G., Janowiak, J. and Xie, P. (1999). GPCP analyses based on observations as a basis for NWP and climate model verification. In *Proceedings of the WCRP 2nd International Reanalysis Conference*, Reading, UK. WCRP report.

Strangeways, I. C. (2003). *Measuring the Natural Environment*, 2nd edn. Cambridge: Cambridge University Press.

Thiessen, A. H. (1911). Precipitation averages for large areas. *Monthly Weather Review*, **39**, 1082–1084.

Todd, M. C., Kidd, C. K., Kniveton, D. R. and Bellerby, T. J. (2001). A combined microwave and infrared technique for estimation of small scale rainfall. *Journal of Atmosphere and Ocean Technologies*, **18**, 742–755.

World Meteorological Organization (1996). *Guide to Meteorological Instruments and Methods Of Observation*, 6th edn. WMO, No. 8. Geneva: WMO.

Xie, P. and Arkin, P. A. (1996a). Analyses of global monthly precipitation using gauge observations, satellite estimates and numerical model predictions. *Journal of Climate*, **9**, 840–858.

Xie, P. and Arkin, P. A. (1996b). An intercomparison of gauge observations and satellite estimates of monthly precipitation. *International Journal of Climatology*, **34**, 1143–1160.

Xie, P., Rudolf, B., Schneider, U. and Arkin, P. A. (1996). Gauge-based monthly analysis of global land precipitation from 1971 to 1994. *Journal of Geophysical Research – Atmosphere*, **101**, 19023–19034.

Yang, D. Q., Kane, D., Zhang, Z. P., Legates, D. and Goodison, B. (2005). Bias corrections of long-term (1973–2004) daily precipitation data over the northern regions. *Geophysical Research Letters*, **32**, L19501.

13

Precipitation means and trends

Limitations of the datasets

If we are interested in how precipitation has changed over the long term we have to rely entirely on raingauge measurements because only they provide a long enough run of data. It will be clear from the foregoing chapters, however, that raingauges are installed mostly on the land with just a few operating on islands to cover the oceans. This means that the greater proportion (71%) of global precipitation is not measured. Even on land, coverage is patchy, with large areas without raingauges, resulting in only about 25% of the earth's surface being covered.

In addition it will be clear from Chapter 8 that even the raingauge data that we do have will be error-prone due to raingauge shortcomings, in particular wind losses, and the situation is worse for snowfall because of increased aerodynamic errors. Finally, conversion from point measurements to areal averages in cells of $0.5° \times 0.5°$, $1° \times 1°$ or $2.5° \times 2.5°$ latitude and longitude also introduces some uncertainty. Since around 1980 satellite measurements have become available and this is improving geographical cover, especially over the oceans and remote land regions. But 25 years is not sufficient to detect long-term trends. In addition, the accuracy of remotely sensed precipitation is still difficult to quantify, and because the better instruments are on polar-orbiting satellites temporal cover is still limited, giving just a brief glimpse of rain events as the spacecraft passes by a few times a day. Despite the low orbits (which cause the low temporal cover) there is still only relatively low spatial resolution, in the order of kilometres, not metres. If we are interested in changes over the last few decades, satellites help, but any trends detected over such a short time could simply represent decadal variability, not long-term trends.

This may sound somewhat pessimistic but the limitations need to be acknowledged so that the findings are seen in context. Nevertheless much can still be deduced.

Global mean annual precipitation

Many attempts have been made to estimate mean global precipitation. Hulme (1995) lists 20 estimates made between 1960 and 1990, the smallest being 784 mm (made in 1964), the largest 1041 mm (in 1975). When adjustments (however imperfect) were made to combat the negative bias of raingauges, the higher estimates of 1130 mm (made in 1978) and 1123 mm (in 1990) resulted.

To give some indication of the way in which precipitation has changed over the last century on the global, hemispheric and regional scales, I am using the data published by New *et al.* (2000, 2001). These are in good agreement with those of Bradley *et al.* (1987), Diaz *et al.* (1989), Vinnikov *et al.* (1990), Eischeid *et al.* (1991), Hulme (1994, 1995) and Dai *et al.* (1997). While the values may differ somewhat between these different estimates, this will be because of the different ways of arriving at gridded values and the different selections of raingauge data used. There is also the difference introduced by the run of data, varying between authors depending on the time the papers were written, with some datasets ending in the 1980s or 1990s and some in the 2000s. For example, by including the 1990s in the analysis, which had lower precipitation, the century-long estimates are reduced. For this reason I have used the New *et al.* data, since they are the most recent (at the time of writing).

In the papers cited above, the authors have presented their data in graphical form, showing precipitation deviations from long-term means, some using millimetre anomalies, some percentage differences, and some standard deviations.

Twentieth-century precipitation trends over land

Global precipitation

The mean annual precipitation for the whole globe over land surfaces during the twentieth century was 1002 mm (New *et al.* 2001), with a positive upward trend of around 8.9 mm or 0.9% (Fig. 13.1A, Fig. 13.2 and Table 13.1). The trend, however, was not a steady linear rise over the century but an increase of about 40 mm from 1901 to 1955, staying above the mean until the 1970s, followed by an equal decline to 1992, since when precipitation has returned back to the century mean.

As New *et al.* (2001) say, 'In contrast to the global temperature record (Jones *et al.* 1999), it is difficult to argue that there has been a significant trend in global land precipitation through the twentieth century.' Hulme (1995) compared land mean precipitation and global mean temperature during the twentieth century and found that there was considerable variability in the relationship, with periods of warming occurring with no change or even a slight reduction in precipitation, while times of increased precipitation occurred during times of little global warming.

Precipitation over land

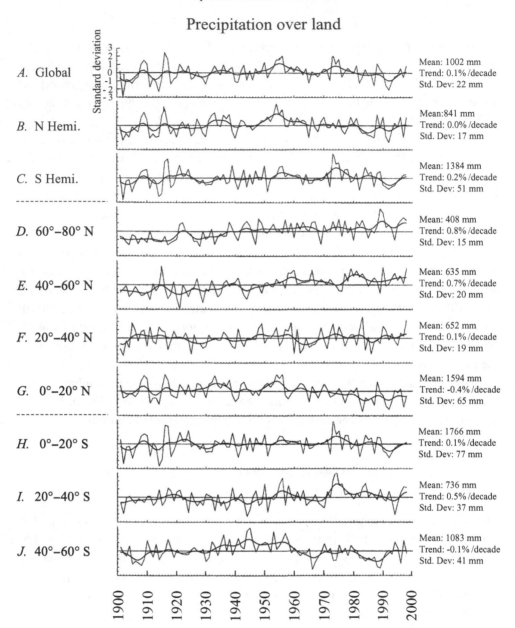

Figure 13.1   Trends in global mean annual precipitation over land during the twentieth century (1901–1998). See text for discussion. From New *et al.* (2001). Copyright Royal Meteorological Society. Reproduced with the permission of John Wiley & Sons on behalf of the RMetS.

Table 13.1 *Precipitation trends over land, 1901–1998 (From New et al. 2001).*

Latitude	Mean (mm)	Change (mm)	Change (%)	Std Dev (mm)
80–90 N	—	—	—	—
60–80 N	408	32.1	7.9	15
40–60 N	635	41.6	6.6	20
20–40 N	652	6.9	1.0	19
0–20 N	1594	−62.8	−4.1	65
0–20 S	1766	18.2	1.0	77
20–40 S	736	36.1	4.9	37
40–60 S	1083	−11.9	−1.1	41
60–90 S	—	—	—	—
N hemisphere	841	3.2	0.4	17
S hemisphere	1384	22.4	1.6	51
Globe	1002	8.9	0.9	22

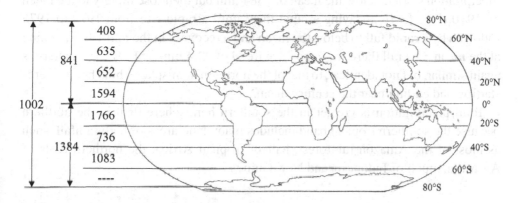

Figure 13.2 The mean annual precipitation over the twentieth century for the whole globe, the two hemispheres and eight latitudinal bands (values based on New *et al.* (2001)).

If North Africa is excluded from the analysis there is a larger increasing trend of 2.2%. Dai *et al.* (1997) justify this exclusion on the grounds that the decline in precipitation after the 1950s over North Africa (which accounts for the falling global trend) is probably a multi-decadal oscillation rather than a long-term trend. This may or may not be the case, and demonstrates the problems of trying to come to a firm conclusion with insufficiently long datasets.

The hemispheres

Because the tropics have the highest rainfall, precipitation in both hemispheres is dominated by that in their tropical belts (0° to 20° north/south). However, since the tropical bands north and south of the equator are different in land distribution and type, they do not behave in exactly the same way, although each dominates its respective hemisphere.

Northern hemisphere

Over the century, the mean annual precipitation over land was 841 mm, slowly rising from below the mean in 1901 to a peak above the mean around 1955 (Fig. 13.1B), thereafter falling back to below the mean in 1980 before returning to the century mean by 2000. The northern hemisphere thus followed the global pattern closely. Despite the quite large interannual and decadal positive and negative variations about the mean, there was only a small overall trend upwards of 3.2 mm or 0.4%.

Southern hemisphere

Precipitation started below the mean of 1384 mm but then rose rapidly to the mean by 1910 (Fig. 13.1C), staying around there until a rapid rise from 1970 to 1975 followed by a rapid fall to below the mean and a recovery in the mid 1990s, ending at the mean. Overall there was a positive trend of 22.4 mm or 1.6%. The pattern is again similar, but not identical, to that of the northern hemisphere, both hemispheres starting and ending near the century mean.

Total precipitation is greater in the southern hemisphere than in the northern because the southern tropical band includes many land areas of high rainfall, such as most of the Amazon rainforest, half of tropical Africa, the northern parts of Australia, much of Indonesia and New Guinea.

Latitudinal zones

Northern tropics (0° to 20° north)

Until peaking in the mid 1950s precipitation stayed close to and slightly above the mean of 1594 mm (Fig. 13.1G), followed by a significant decrease until the mid 1980s, whereafter the trend levelled out but did not return to the mean. Overall there was a downward trend of 62.8 mm or –4.1%, due mostly to the Sahelian drought.

Northern subtropics (20° to 40° north)

There was little overall trend in these latitudes throughout the century (Fig. 13.1F), precipitation year by year remaining close to the century mean of 652 mm, the overall upward trend being 6.9 mm or 1%.

Northern mid latitudes (40° to 60° north)

Starting below the century mean of 635 mm, precipitation stayed below the mean for the first half of the century (Fig. 13.1E), rising above the mean around 1950 and remaining there for the next 50 years, the overall trend being 41.6 mm, an increase of 6.6%. The probable reason for this (apparent?) increasing trend is explained in the next paragraph.

Northern high latitudes (60° to 80° north)

This band behaved like its southerly neighbour, 40–60° north, starting below the century mean of 408 mm (Fig. 13.1D), reaching the mean in 1940, continuing to rise slowly until at the end of the century it had risen 32.1 mm, an increase of 7.9%. This increase, and possibly also the increase in the 40–60° belt may, in part at least, be due to the same amount of precipitation falling but with more of it falling as rain and less as snow, due to a warming climate (Førland and Hanssen-Bauer 2000). Since raingauges catch rain more efficiently than snow the impression might be wrongly gained that precipitation had increased. The increase is mostly in spring, reinforcing the view that a greater proportion fell as rain. But there is also the matter of the former Soviet Union introducing Tretyakov screens from the 1950s, which improve the catch of snowfall, so the trend may be an artefact of the raingauge network.

Southern tropics (0° to 20° south)

There was little overall precipitation change over land during the century (Fig. 13.1H), with an increase of 18.2 mm or 1.0%, starting and ending near the mean of 1766 mm.

Southern subtropics (20° to 40° south)

This predominantly desert band had a mean precipitation of 736 mm but remained below this until the 1950s (Fig. 13.1I). Due partly to a wet period through the 1970s, the subtropics experienced an increase of 36.1 mm or 4.9%, but ended only marginally above the average.

Southern mid latitudes (40° to 60° south)

Precipitation over the land areas of this band had a century mean of 1083 mm. There was no overall strong trend but the means varied quite considerably year by year (Fig. 13.1J), being below the century mean until 1930, becoming positive until 1960 and thereafter falling back below the mean again, returning to the mean in the last few years of the twentieth century. The small overall trend was downwards, ending 11.9 mm below its starting point, or −1.1%.

But this band contains very little land indeed, represented only by Patagonia, the South Island of New Zealand and Tasmania (Fig. 13.2). In consequence, errors could be very large since the gauges may be in unrepresentative locations, compounding the problem that land covers only a very small fraction of the area.

These values are summarised in Table 13.1 and Figure 13.2.

Precipitation trends over the oceans

Tropical islands

Detecting trends over the oceans is difficult because of the sparsity of raingauge measurements, gauges mostly being operated on islands. Nevertheless, because oceans are more homogeneous than the land, we might expect that precipitation would be more uniform over the oceans, and this is confirmed by the observed trends at the tropical islands, which show consistency over large parts of the oceans. The values given below are for a band from 30° south to 30° north, and for a slightly shorter period than for the land, starting in 1920 rather than 1901. Again they are from New *et al.* (2001).

Pacific Ocean

Over the *western Pacific islands* precipitation undulated around the 80-year average of 2933 mm (Fig. 13.3A) until a fall to below the mean started around 1960; it then remained at this lower level, apart from a slight temporary increase around 1980, the negative trend being 91.8 mm over the full period or −3.1%.

The *eastern Pacific islands* experienced a gradual decrease in precipitation from 1920 to 1950 (Fig. 13.3B), after which there was an increase to above the 80-year mean of 1877 mm by 1960, falling again to below the average by the mid 1970s and then a large rise that continued during 1975–1985, followed by a fall to below the mean near the close of the century, the overall upward trend being 31 mm over the 80 years or 1.7%.

The coincident positive and negative trends in the east and west Pacific from about 1975 are probably due to increased El Niño activity in the second half of the twentieth century (see next chapter).

Atlantic Ocean

There are few data from the Atlantic but from those that are available there appears to have been an increase in precipitation in the south and a decrease in the north (in the tropical belts of 30° south to 30° north). Combining north and south, the trend was upwards from below the 80-year mean of 1444 mm at the start (Fig. 13.3C) to above the mean by 1928, remaining positive until falling below the mean again

Precipitation over oceanic islands
(30°S –30°N)

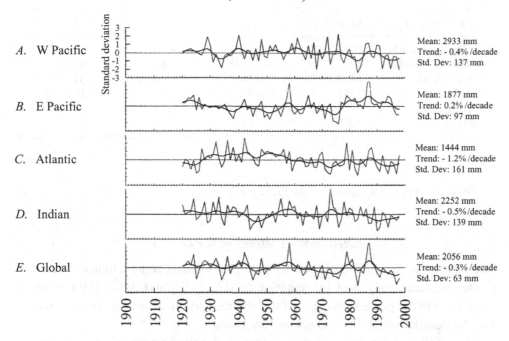

	Standard deviation		
A. W Pacific			Mean: 2933 mm Trend: - 0.4% /decade Std. Dev: 137 mm
B. E Pacific			Mean: 1877 mm Trend: 0.2% /decade Std. Dev: 97 mm
C. Atlantic			Mean: 1444 mm Trend: - 1.2% /decade Std. Dev: 161 mm
D. Indian			Mean: 2252 mm Trend: - 0.5%/decade Std. Dev: 139 mm
E. Global			Mean: 2056 mm Trend: - 0.3% /decade Std. Dev: 63 mm

Figure 13.3   Trends in global mean precipitation over the oceans in the band 30° S to 30° N during the twentieth century from 1920, based on measurements made by raingauges on islands. See text for discussion. From New *et al.* (2001). Copyright Royal Meteorological Society. Reproduced with the permission of John Wiley & Sons on behalf of the RMetS.

around 1960, where it remained for the rest of the century. The overall trend was down 134 mm or −9.3%.

Indian Ocean

Precipitation reduced over the western side of the ocean while it rose over the east (Fig. 13.3D), the overall mean being 2252 mm with a negative trend of 97.2 mm over the 80 years or −4.3%.

Global islands

Combining all of the island data (Fig. 13.3E), which can be taken as a rough proxy for precipitation over all of the tropical oceans (30° south to 30° north), the mean for the 80 years was 2056 mm, with a decline of 51.36 mm over the 80 years or −2.5%. Since tropical rainfall dominates, this will be a major factor in global precipitation, although due to a sparsity of gauges, it is only approximate.

Table 13.2 *Precipitation trends over the oceans 1920–1998*
(From New et al. 2001).

Location 30 S–30 N	Mean (mm)	Change (mm)	Change (%)	Std Dev (mm)
West Pacific islands	2933	−91.8	−3.1	137
East Pacific islands	1877	31.0	1.7	97
Atlantic islands	1444	−134.0	−9.3	161
Indian Ocean islands	2252	−97.2	−4.3	139
Global islands	2056	−51.4	−2.5	63

These values are summarised in Table 13.2.

Satellite observations of the oceans

The largest satellite record of monthly precipitation data is the Climate Prediction Center Merged Analysis of Precipitation (CMAP) (Xie and Arkin 1996) running from 1979 to the present. Because of the shortness of the span any apparent trends may be decadal variability and not secular changes.

The satellite data suggest that precipitation decreased over much of the ocean from 10° to 30° south except for the eastern boundaries of the Pacific and Atlantic oceans, where it increased. In the northern hemisphere oceans, there was a similar but smaller decrease. Over the narrow tropical band 5° north to 5° south across much of the Pacific and Atlantic there was increased precipitation. Beyond the tropics over the southern oceans there were alternate bands of increasing (30–45° S), decreasing (45–60° S) and increasing (60–75° S) precipitation. The predominantly downward trend in satellite observations supports the negative trend measured by raingauges on the global islands (Fig. 13.3E). But this is still provisional.

Precipitation trends since 1766 over England and Wales

Some parts of the world have had a good raingauge network since the eighteenth century, the UK being the leading example and the pioneer of such measurements. Although they do not throw light on global changes, it is interesting to look at what changes there have been in precipitation over the last 240 years in the UK.

Figure 13.4 illustrates the trends in precipitation since 1766, probably the longest record available from anywhere, and it can be seen that there has been little change over that period. The mean annual precipitation (1766–2005) was 914.19 mm with

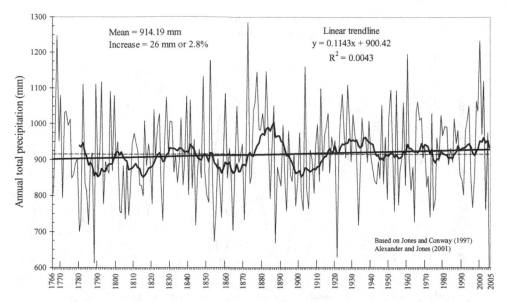

Figure 13.4 Annual mean precipitation in the UK from 1766 to 2005 shows a slight linear upward trend of about 2.8%. See text for discussion. Based on Jones and Conway (1997) and Alexander and Jones (2001). Crown Copyright. Data provided by the Met Office.

a small positive linear trend of 26 mm over the whole period, an increase of 2.8% (or 0.12% per decade over the 240 years). This is in reasonable accord with the global picture (Fig. 13.1A), in both cases there being only a small increase in precipitation, both ending near the mean, strengthening the view that precipitation has been quite stable across the globe since records began. The little ice age was only just past its minimum in the mid 1700s, temperatures rising steadily from their 1600–1700 low to the present, yet precipitation has not changed significantly in the UK over that period.

Although total annual precipitation changed little between 1766 and 2005 in the UK, there was a notable seasonal shift (Fig. 13.5). When plotted as two curves, high-summer (July and August) and winter (December to March), summer precipitation is seen to have fallen while winter precipitation increased, both by around 20% over the period, particularly after about 1870. These changes are thought to be due to natural changes in atmospheric circulation.

Summary

It is useful in summary to return to the global view, for the detailed changes of hemispheres and zones do not leave a clear mental picture. The mean precipitation

Seasonal England and Wales Precipitation
High Summer and Winter, 1766–2005

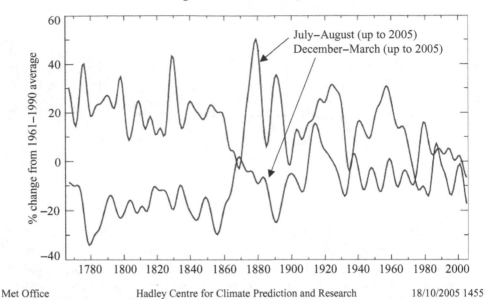

Met Office   Hadley Centre for Climate Prediction and Research   18/10/2005 1455

Figure 13.5 Annual precipitation in the UK from 1766 to 2005 shows a decreasing trend in summer and an increasing trend in winter of about 20%. Most of the changes date from around 1870 onwards. See text for discussion. Crown Copyright. Graph provided by the Met Office.

for the whole globe over land during the twentieth century was about 1000 mm with an overall upward trend of 8.9 mm or 0.9%, with peaks and troughs throughout the century on each side of the long-term mean, ending at the close of the century back at the century mean. There is, therefore, no significant trend in annual total precipitation over land to match the increase in global temperature.

Over the UK, the longer-duration raingauge network has also shown that there has been only a small trend in the UK over the last 240 years with a rise of 2.8%. There has, however, been a gradual shift in the seasonal distribution from the 1870s onwards, with drier summers and wetter winters most probably due to natural changes in atmospheric circulation.

Over the oceans we can be rather less certain as to changes, but taking the readings for the whole globe from islands in the band 30° south to 30° north, there was a decline of precipitation, punctuated by decadal-length undulations, with an overall fall of 2.5%.

Trends and means are the not the whole story, however. Variability and extremes are also important, and the next chapter investigates these.

References

Alexander, L. V. and Jones, P. D. (2001). Updated precipitation series for the UK and discussion of recent extremes. *Atmospheric Science Letters*, **1**, 142–150.

Bradley, R. S., Diaz, H. F., Eischeid, J. K., Jones, P. D., Kelly, P. M. and Goodess, C. M. (1987). Precipitation fluctuations over Northern Hemisphere land areas since the mid-19th century. *Science*, **237**, 171–175.

Dai, A., Fung, I. and Del Genio, A. (1997). Surface observed global land precipitation variations during 1900–88. *Journal of Climate*, **10**, 2943–2962.

Diaz, H. F., Bradley, R. S. and Eischeid, J. K. (1989). Precipitation fluctuations over global land areas since the late 1800s. *Journal of Geophysical Research – Atmosphere*, **94**, 1195–1210.

Eischeid, J. K., Bradley, R. S. and Jones, P. D. (1991). *A Comprehensive Precipitation Data Set for Land Areas*. DOE Technical Report TR051. DOE ER-69017T-H1. Available from National Technical Information Service, US Department of Commerce.

Førland, E. J. and Hanssen-Bauer, I. (2000). Increased precipitation in the Norwegian arctic: true or false? *Climate Change*, **46**, 485–509.

Hulme, M. (1994) Validation of large-scale precipitation fields in general circulation models. In *Global Precipitation and Climate Change* (ed. M. Desbois and F. Desalamand). Berlin: Springer, pp. 387–406.

Hulme, M. (1995). Estimating global changes in precipitation. *Weather*, **50**, 34–42.

Jones, P. D. and Conway, D. (1997). Precipitation in the British Isles: an analysis of area averaged data updated to 1995. *International Journal of Climatology*, **17**, 427–438.

Jones, P. D., New, M., Parker, D. E., Martin, S. and Rigor, J. G. (1999). Surface air temperature and its changes over the past 150 years. *Reviews of Geophysics*, **37**, 173–199.

New, M., Hulme, M. and Jones, P. D. (2000). Representing twentieth century space-time climate variability. Part II: development of 1901–1996 monthly grids of terrestrial surface climate. *Journal of Climate*, **13**, 2217–2238.

New, M., Todd, M., Hulme, M. and Jones, P. (2001) Precipitation measurements and trends in the twentieth century. *International Journal of Climatology*, **21**, 1899–1922.

Vinnikov, K. Y., Grolsman, P. Y. and Ligina, K. M. (1990). Empirical data on contemporary global climate changes (temperature and precipitation). *Journal of Climate*, **3**, 662–677.

Xie, P. and Arkin, P. A. (1996). Analyses of global monthly precipitation using gauge observations, satellite estimates and numerical model predictions. *Journal of Climate*, **9**, 840–858.

14

Precipitation variability and extremes

Behind the long-term means and trends, discussed in the previous chapter, there is a much greater short-term variability of precipitation year by year at all geographical scales. Precipitation can also vary in its extremeness, for example shifts to greater or lesser intensity of individual storms or to changes in the length and frequency of wet and dry periods. We will look at these two aspects of precipitation separately, starting with variability.

It has been shown by many researchers that the variability of precipitation in the tropics (and also further afield), is highly influenced by the cyclic variations of sea surface temperature (SST) across the equatorial Pacific caused by El Niño. We need first to look at these processes and then at the effects they have on precipitation.

El Niño and La Niña

Every year, about Christmas time, a warm southerly-flowing current originating from around the Galapagos Islands replaces the very cold northward current along the coasts of Peru and Ecuador. Every few years the southern current is much warmer. This has been known for centuries by the local inhabitants, but it was only recently recognised that these periodic events extended over the entire tropical Pacific (Bjerknes 1969, Allen *et al.* 1996, *Weather* 1998). While originally the term *El Niño* was used by fishermen to describe the local annual warm current (it means 'the Christ child' in Spanish, after its time of onset) it is now used for the large-scale warmer events across the entire Pacific. More recently the term La Niña (the girl child) was introduced (Philander 1990), describing the opposite conditions to El Niño.

As Fig. 14.1 illustrates, the ocean is composed of two layers – a warm very shallow top layer about 100 metres deep above a cold layer extending down to depths of 4000 metres or more, the two layers being separated by the *thermocline*, a thin layer across which the temperature gradient is high.

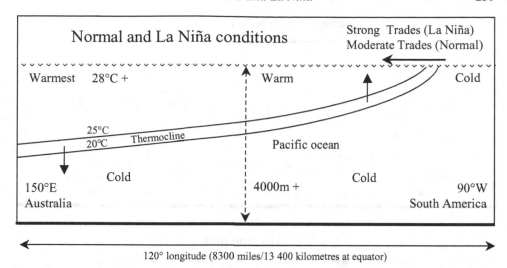

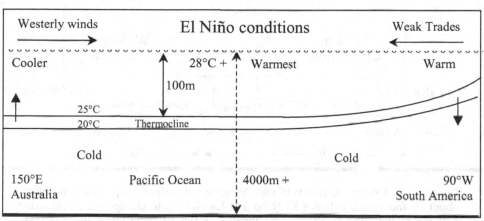

Figure 14.1 El Niño and La Niña are the names given to the two different states of the surface waters of the equatorial Pacific. La Niña can be considered as a more extreme form of the 'normal' conditions, the strong trade winds holding the warmer surface water to the west and allowing cooler water to rise to the surface in the east and central Pacific. El Niño is the opposite state caused by the trade winds weakening, allowing warmer water to cover the entire Pacific. The causes of these alternating states and their effects are discussed in the text.

Normal conditions

'Normal' conditions are simply a mild form of La Niña and, as Fig. 14.1 (upper) shows, moderate trade winds force the warm surface water of the Pacific westward, causing the thermocline to rise towards the surface in the east, where it reduces the sea surface temperature compared with that in the west. Due to the higher surface temperature of the ocean in the west, convective storms occur over Indonesia and Australia, with heavy rainfall and thunder. The thermals rise to near the tropopause,

Normal and La Niña conditions

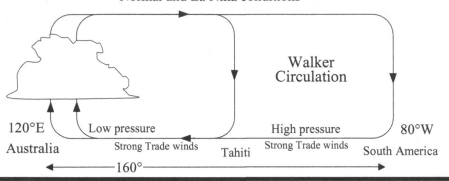

El Niño conditions

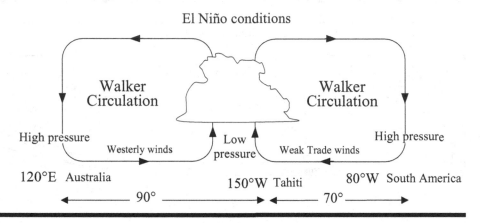

Figure 14.2 One of the effects of the changing sea surface temperatures brought about by the switch between El Niño and La Niña is to change the circulation patterns of the atmosphere over the Pacific. In the La Niña phase there is one single large Walker Circulation producing higher pressure in the east and lower pressure over Australia with resultant cumulus cloud formation and precipitation. In the El Niño phase the convective activity occurs more over the central regions of the now warmer Pacific, producing two Walker Circulations with a resultant low pressure over Tahiti and higher pressure over Australia and South America. This figure should be looked at in combination with Fig. 14.1, for the processes are interrelated.

resulting in upper-level westerly winds that travel in a closed 'Walker Circulation' as far as South America, where they subside, causing high barometric pressure and drought, returning to the west as moderate trade winds (Fig. 14.2, upper).

La Niña

This condition is just a more extreme form of the 'normal', in which the trade winds are more intense, forcing the warmer waters still further west, raising the thermocline in the east higher, causing greater cooling of the sea surface (Fig. 14.1, upper); the wind and precipitation patterns are retained but intensified.

El Niño

When some small perturbation in the trade winds causes them to lessen slightly, the warm water starts to move back eastward and this enhances the reduction of the trade winds, which causes a still further eastward movement of the warm surface, combined with a sinking of the thermocline as it levels off to become almost horizontal. El Niño is now established (Fig. 14.1, lower), the warm sea surface producing large cumulus clouds with high rainfall over the central parts of the Pacific. There are now two Walker Circulations (Fig. 14.2, lower), the much warmer central Pacific producing a shift of the precipitating cumulonimbus clouds eastward from over Australia, accompanied by low barometric pressure, Tahiti being in the midst of this new rainfall activity while the barometric pressure over Australia rises due to the descending air, the city of Darwin being near its centre. Over to the east in South America, the descending Walker Circulation produces a comparable high pressure. Sooner or later, however, the trades pick up again and force the warm water back to the west and La Niña returns.

What causes these changes in wind that result in the oscillation of barometric pressure from alternately low in Australia and high in Tahiti during La Niña to the opposite state of affairs during El Niño? Although the changes in sea surface temperature cause the changes in the atmosphere, in the form of wind, precipitation and barometric pressure modifications, the atmosphere also causes the changes in the sea surface temperature. So which leads? What causes the eastward movement of the warm surface waters that result in a switch from La Niña to El Niño?

The atmosphere and ocean are not equals in this and are not balanced. The atmosphere is lightweight, changing quickly in response to changes over the ocean in a matter of days or weeks. The ocean, on the other hand, having a high inertia, takes months to adjust to wind changes. So the ocean state at any one time is not determined simply by the present winds, since it is still responding and adjusting to earlier winds. It has, in effect, a memory in the form of waves. The waves act below the surface, moving the thermocline up in one place and down in another, and this affects the transition from one phase to the next. This makes it important to measure the ocean conditions so as to be able to anticipate future changes (see 'Measuring the sea surface temperatures', below).

Each phase typically lasts about 9–12 months, usually beginning to form during June to August, reaching a peak from December to April, when it has most effect on precipitation, and decaying away during May to July of the year following.

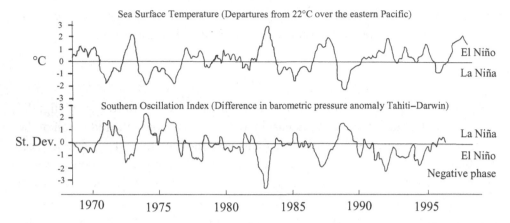

Figure 14.3 The combined process of sea surface temperature (SST) and barometric pressure changes over the Pacific are known collectively as the El Niño Southern
Oscillation (ENSO). These two graphs demonstrate the close synchronism of the
departures of the SST (top curve) and departures of barometric pressure (lower
curve) as El Niño changes into La Niña and back. The graphs are based on data
from the NOAA Climate Prediction Center website.

Some episodes can, however, last up to two years, sometimes even longer, and their
periodicity is quite irregular, although on average El Niño and La Niña occur every
3–5 years.

The Southern Oscillation

From the above it will be seen that when La Niña is active, the barometric pressure is high in the central and eastern Pacific and lower in the west. When El
Niño is operative the pressure is low in the central Pacific and high over the
west and east. This seesawing of barometric pressure was named the *Southern
Oscillation* by Walker and Bliss (1932). Pressures are measured at Darwin in
Australia and Tahiti in the central Pacific, and the difference (Tahiti – Darwin)
taken. During an El Niño event, the difference is negative (and hence is known
as the negative phase). The combined process of changes in the SST and barometric pressure are known collectively as the *El Niño Southern Oscillation* or
ENSO.

　　If the SST variations, as departures from 22 °C, in the central and eastern Pacific
are plotted, the cyclic changes can be seen to coincide with the plotted difference
in pressure between Tahiti and Darwin (Fig. 14.3).

　　It is useful to be able to quantify the Southern Oscillation and for this purpose
the concept of the Southern Oscillation Index (SOI) was created. There are several

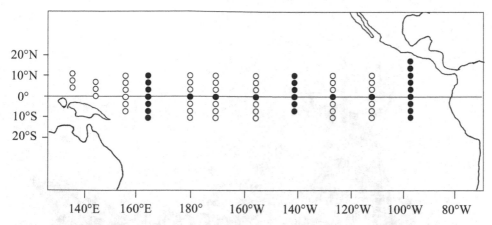

Figure 14.4 To monitor the sea surface temperatures of the equatorial Pacific and to follow the changes between El Niño and La Niña, an array of moored buoys has been installed across the tropical Pacific at roughly the positions shown, the project being known as the Tropical Atmosphere Ocean/Triangle Trans-Ocean Buoy Network (TAO/TRITON). The buoys shown as dark discs measure precipitation, using the capacitive type of raingauge described in Chapter 8 (Serra *et al.* 2001).

ways of calculating this (Ropelewski and Jones 1987, Konnen *et al.* 1998), the method used by the Australian Weather Service being as follows:

$$SOI = 10\,(P_{diff} - P_{diffav})/\mathrm{SD}(P_{diff})$$

where P_{diff} is (average Tahiti mean-sea-level pressure for the month) – (average Darwin mean-sea-level pressure for the month)

P_{diffav} is the long-term average of P_{diff}

$\mathrm{SD}(P_{diff})$ is the long-term standard deviation of P_{diff} for the month

Measuring the sea surface temperatures

With an increasing understanding of the workings of ENSO it became useful to have measurements of SST as well as the variations in barometric pressure (which are quite easy to measure, needing only to be measured in the two places). To measure SST, an array of moored buoys has been installed across the full width of the tropical Pacific between 10° north and 10° south, the project being known as the Tropical Atmosphere Ocean/Triangle Trans-Ocean Buoy Network (TAO/TRITON) (Fig. 14.4) (Serra *et al.* 2001). The equipment was designed and made at NOAA's Pacific Marine Environmental Laboratory (PMEL), supported by Japan and France, the buoys forming the Autonomous Temperature Line Acquisition System (ATLAS). Started in 1985 and completed in 1994, there are currently about 70 buoys telemetering data back to shore in real time via satellites. The buoys

Figure 14.5 The UK Met Office operates nine deep-water buoys moored on the edge of the continental shelf in the northeastern Atlantic; the one illustrated is receiving a six-monthly service visit. This is similar to the TAO/TRITON buoys except that precipitation is not measured. Photograph reproduced by permission of the UK Met Office.

measure more than just SST, including windspeed and wind direction, air temperature, humidity, incoming solar radiation, incoming long-wave radiation, barometric pressure, sub-sea surface temperature, salinity and ocean currents. About 27 of the buoys also measure precipitation using the capacitive type of raingauge (described in Chapter 8), but without any arrangement to keep the gauges level (see Chapter 8 for an explanation of this).

There are also 12 similar buoys, all measuring precipitation, known as the Pilot Research Moored Array in the Tropical Atlantic (PIRATA). Similar moored buoys are also operated by the UK Met Office in the Atlantic on the edge of the continental shelf, and one of these is illustrated in Fig. 14.5. They do not, however, measure precipitation.

For additional background, readers are referred to *Weather* (1998), which is devoted entirely to ENSO. There are also many web sites, which can be found by doing a search under 'ENSO' or 'El Niño', and those of NOAA are particularly useful.

The influence of ENSO on precipitation

Since we are looking at precipitation variability, it is more informative to look not at rainfall totals but at how the precipitation changes when switching from La Niña to El Niño episodes. The variations are best shown diagrammatically (Figs. 14.6, 14.7). The patterns of change that I am using here are from NOAA's Climate Prediction Center, which I have redrawn to show just precipitation (the originals contain temperature data as well).

El Niño events cause climate anomalies to occur in the immediate neighbourhood, over the equatorial Pacific where El Niño occurs, and also well beyond – by 'teleconnections' (through upper atmosphere processes). The following sums up the figures. The boundaries shown are only approximate.

El Niño during December to February

Wetter conditions

During its peak in December to February, the movement of warm surface water eastward across the Pacific shifts convective activity from its normal position around Australia towards the central and eastern Pacific, increasing precipitation in these regions (Fig. 14.6, upper), extending as far as the coast of Ecuador and northern Peru. The effect continues southeastward across South America to southern Brazil and central Argentina, northwards to the southern USA and westward to equatorial east Africa.

Drier conditions

At the same time, precipitation reduces over Australia, Indonesia, New Guinea, Malaysia, the Philippines and northern South America. The effect is also felt to the west over the Indian Ocean, causing the monsoon rains to fail in strong El Niños, continuing as far west as southeast Africa and Madagascar, where rainfall is lower than normal.

El Niño during June to August

Wetter conditions

The effects of El Niño are less during these months, with increased precipitation being limited to the central Pacific equatorial region, to a small region in central Chile and over the intermountain region of the US northwest (Fig. 14.6, lower).

Drier conditions

The drier than usual conditions occurring in the December to February period continue over Australia, Indonesia, New Guinea, Malaysia and India but extend further south across eastern Australia and as far as the North Island of New Zealand (Fig. 14.6, lower).

El Niño precipitation anomalies
December - February

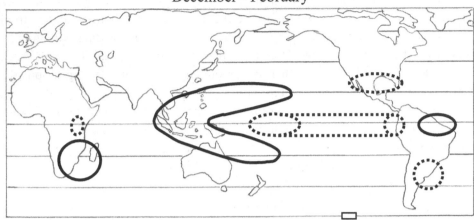

June - August

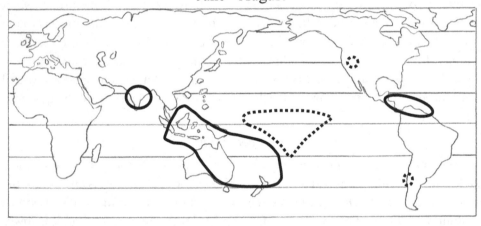

——————— Drier conditions
••••••••••••••• Wetter conditions

Figure 14.6 The way in which El Niño affects precipitation is illustrated here, the upper map showing the anomalies during December to February – its most active time – while the lower map shows conditions during June to August. These maps are based on those published by NOAA's Climate Prediction Center on their website (www.cpc.noaa.gov).

La Niña during December to February

Wetter conditions

Since La Niña is an enhanced 'normal' condition, precipitation departures tend to be more or less the exact opposite of those during El Niño (Fig. 14.7, upper).

La Niña precipitation anomalies
December - February

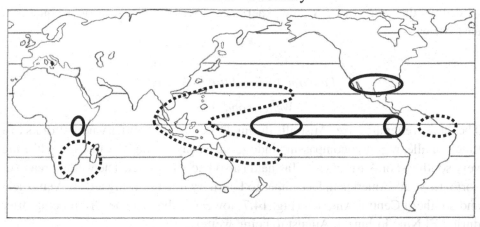

June - August

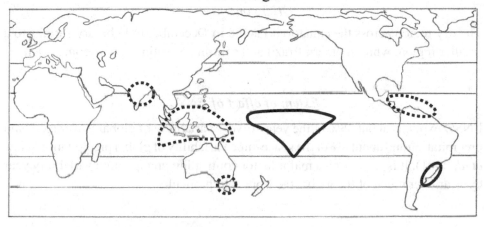

————— Drier conditions
·············· Wetter conditions

Figure 14.7 These are the counterpart maps for the La Niña phase. It is useful to compare the conditions here with those shown in Fig. 14.6 for El Niño; they are almost perfect opposites. These maps are based on those published by NOAA's Climate Prediction Center on their website (www.cpc.noaa.gov).

Rainfall is now greater across the western Pacific, with higher rainfall over Australia, Indonesia, New Guinea, Malaysia and the Philippines. The dry conditions over the north of South America and southeast Africa during El Niño now switch to being wetter.

Drier conditions

The central and eastern equatorial Pacific and the coast of Ecuador and northern Peru as well as the southern USA now experience drier conditions, as does equatorial east Africa.

La Niña during June to August

Wetter conditions

The large wetter area over Australia during December to February now shrinks to a much smaller region encompassing Indonesia, Malaysia and New Guinea, while the very southeast of Australia and Tasmania also become wetter. The wetter area over northern South America in December to February moves northwards to Venezuela and southern Central America (Fig. 14.7, lower). India switches from being drier during El Niño in June to August to being wetter.

Drier conditions

The dry region across the equatorial Pacific in December to February shrinks to a smaller region, while southern Brazil and central Argentina now become drier.

Extent of effect of ENSO

ENSO explains about 38% of the year-to-year variability of globally averaged land precipitation and about 8% of the space-time variability of global precipitation (New *et al.* 2001). It is, therefore, a major factor controlling precipitation variability year by year over much of the globe, the tropics in particular.

El Niño's effect on tropical cyclones

Perhaps not surprisingly, the warmer than usual SSTs across the Pacific during El Niño affect tropical cyclone activity, their frequency increasing in the central and eastern Pacific, while north of the equator to the west there is a reduction in the number of tropical storms (Lander 1994). Hurricanes also tend to originate slightly closer to the equator during an El Niño event.

The North Atlantic, Arctic and Antarctic oscillations

In addition to the Southern Oscillation, there are several other similar pressure oscillations, although lesser in the sense that they do not have the same strong global influence, in part because they are not in the equatorial zone where most precipitation occurs. Nevertheless they do account for some of the variability of climate, including precipitation, in the higher and mid latitudes.

The North Atlantic Oscillation (NAO)

The NAO is the main cause of winter precipitation variability (and climate generally) in the North Atlantic region covering central North America to Europe, extending into Northern Asia. The NAO is the seasaw of barometric pressure differences between the Azores subtropical high and the Icelandic polar low (Jones *et al.* 1997). The NAO index is defined as the anomalous difference between the polar low and the subtropical high. It is most active during the winter season (December to March) and varies from year to year, although it can remain in one state for several years.

In the 'positive' phase there is a stronger than usual Azores high pressure and a deeper than usual Icelandic low. The increase in pressure gradient causes more, and more intense, winter depressions to cross the Atlantic on a more northerly trail, causing warm wet winters in Europe, wet conditions from Iceland to Scandinavia, drier conditions in southern Europe, cold dry winters in northern Canada and Greenland, while the eastern USA has mild wet winters. These patterns of weather have been known since Walker and Bliss (1932) discovered them.

A negative pressure gradient occurs when the subtropical high and the polar low are both weaker than usual. The reduced pressure gradient results in fewer and weaker depressions, which also take a more west-to-east route (rather than the more usual northeast route – Chapter 4).

Over the past 30 years the NAO has tended towards a more positive phase, which appears to be unusual compared with the last few centuries (Hurrell 1995), being more pronounced since 1989. This behaviour has added to the debate concerning our ability to detect or distinguish between natural and anthropogenic climate change, Hurrell and van Loon (1997) for example showing that the recent NAO trend accounts for much of the observed regional surface warming over Europe and Asia. However, Corti *et al.* (1999) suggest that anthropogenic climate change might influence modes of natural variability. These are central research questions but they do illustrate the complexity of climate processes and how they are still incompletely understood.

At present there is no consensus on what mechanisms cause the observed variations in the NAO, although the Hadley Centre at the UK Met Office uses a forecasting method based on the distribution of SST anomalies across the western North Atlantic during the preceding May. The UK Met Office forecast that the winter of 2005/6 might experience a negative-phase NAO and thus a colder winter than we have had over the last decade. This put the news media into a frenzy of

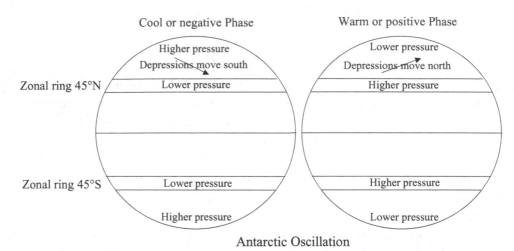

Figure 14.8 The Arctic Oscillation (AO) is the seesaw of barometric pressure between the northern polar and mid latitudes, while the Antarctic Oscillation (AAO) is the counterpart in the southern hemisphere. Their effects on precipitation are discussed in the text.

exaggeration – as they are prone to do over everything – predicting power cuts and other disasters. In the event, while it was a slightly colder winter than of late in the UK, apocalypse failed to materialise. (The media did not come back and say 'sorry about the alarm', they just went on with the next one, or two . . .)

The NAO accounts for about 10% of the variance in land precipitation in the North Atlantic region (Dai *et al.* 1997) and about 8% of annual variability of precipitation averaged over the northern hemisphere's mid latitudes. Its effect is greatest during winter.

The Arctic Oscillation (AO)

The AO is the changing balance of barometric pressure between northern mid and high latitudes (Fig. 14.8). The AO index is said to be negative when there is relatively high pressure over the polar region and lower than normal pressures in the mid latitudes (a zonal ring between 40° and 50° north); the positive phase is when the pressures are reversed.

In the positive or warmer phase, the higher mid-latitude and lower polar pressures cause depressions to move further north, bringing an increase in precipitation to Alaska, Scotland and Scandinavia, along with reduced precipitation to the western USA and the Mediterranean. The negative phase has the opposite effect, bringing drier and colder conditions to the UK and Scandinavia and increased precipitation in the Mediterranean. During most of the twentieth century the AO alternated between the two phases, but starting in the 1970s there has been a tendency for it to stay in the positive phase. This reflects the same tendency noted above for the NAO. Indeed there is a high correlation between the NOA and the AO and there is some debate as to whether the NAO is simply a regional manifestation of the AO.

The AO explains 48% and 35% of winter precipitation variability over land in the latitude bands 60–80° north and 40–60° north respectively (New *et al.* 2001).

The Antarctic Oscillation (AAO)

This is the southern counterpart of the AO, the AAO index being the difference in normalised surface-level pressures between the polar regions and a zonal ring near 45° south (Fig. 14.8).

The AAO appears to have less influence on southern hemisphere precipitation than the AO on the northern, but since mid-latitude precipitation over land in the south includes only southern South America, New Zealand and Tasmania it is unrepresentative of the area generally, which is mostly ocean, from which we have few measurements – which is again the case for the southern high latitudes of Antarctica. When satellite data are included, which include the oceans, the correlation is higher, but the run of data is relatively short, so at the moment the effect of the AAO on precipitation is inconclusive. Because there is more land in the northern hemisphere, there are many more precipitation data from there. In addition there is a much longer run of data because measurements started there earlier. Consequently more is known about the AO effects on precipitation than about the AAO.

Precipitation extremes

So far in this chapter we have considered the causes and nature of precipitation variability on interannual and decadal scales, but precipitation can also vary in its extremes. This last section addresses the question of whether there is any pattern to this.

What is extreme?

Extremes include more intense local convective rain or more intense depressions bringing heavier rain from nimbostratus. Longer dry or wet spells are another expression of extremes, as are the numbers of days with rain. So 'extreme' means a number of different things. In one study the definition of extreme is 'a daily episode that results in two or more inches of precipitation', this being a special case of the more general 'excess over threshold'. But since precipitation totals and intensities vary greatly and naturally around the globe, it is not possible to compare extreme events in different parts of the world. So there can be no 'global mean' of precipitation extremes, no one figure that tells all. Consequently it is more difficult to come to a firm conclusion about whether and how precipitation extremes are changing. It is also unwise to extrapolate from changes in extremes detected over just a few decades in localised parts of the globe to infer what might be happening over longer periods over the whole globe, for short-term changes might be simple decadal variations of a cyclic nature.

Since daily precipitation totals are the most detailed data available, and only from limited regions of the world, most data being in the form of monthly totals, the fine detail of rainfall intensity (in the sense of minute-by-minute rainfall rates) is not accessible. Such observation could be derived from an examination of the chart records from Dines-type recording raingauges, but this would be very labour-intensive. Of more use would be the records from tipping-bucket gauges, but only if the time and date of each tip was logged (rather than a daily total). Such records, however, only became available recently (discounting Sir Christopher Wren's AWS in the seventeenth century). Monthly totals can of course tell us something about long-term extremes – long wet or dry spells. With these caveats in mind, let us look at a few examples of studies of extremes.

Extremes derived from daily data

Karl and Knight (1998) analysed daily precipitation data from 1910 to 1996 over the USA and found that an increase of precipitation of about 10% had occurred, most of the increase being due to an increase in the number of days with rain (6.3 days per century). They also found an increase in the frequency of heavy precipitation.

Groisman *et al.* (1999) analysed daily precipitation data from the USA, Canada, Mexico, the former Soviet Union, China, Australia, Norway and Poland (about 40% of the land surface) and found increases in summer precipitation (except over China) of around 5% during the twentieth century. This is the exact opposite to what has occurred in the UK (Fig. 13.5), where there has been a reduction in summer precipitation, although this could still include more intense rain events on the daily or hourly time scale.

Still looking at daily precipitation, Hennessy *et al.* (1999) found that over much of Australia the annual total precipitation increased by 14–18%, although the changes were of a mixed pattern across the country, with both increases and decreases in both heavy and light rain. Over Australia as a whole, the number of days with rain increased by 10%. (However, see Dai *et al.* 1997, below, on the influence of ENSO on Australian rainfall.)

In Japan, daily precipitation records analysed by Iwashima and Yamamoto (1993) revealed that there were more intense events in recent decades. But as noted earlier we must beware not to extrapolate this backwards or forwards and infer long-term changes based on recent decades.

In the UK, Osborn *et al.* (2000) analysed the frequency of 'heavy' rain (daily precipitation contributing more than 10% of the seasonal precipitation total) at 110 stations over the period 1961–1995. They found that in the winter there was an increase in the contribution from the 'heavy' events while in the summer there was a decrease; in spring and autumn there was no change. This study covered only the second half of the twentieth century but its findings parallel the long-term trend seen in Fig. 13.5 that has been taking place in the UK since the mid nineteenth century, of decreasing precipitation in the summer accompanied by an increase in the winter.

Kunkel *et al.* (2003), using a more recent dataset of daily data from the USA compiled by the National Climatic Data Center (NCDC) covering the period 1895–2000, found that there was a similar level of extreme daily precipitation in the late nineteenth and early twentieth century to that seen in the late twentieth century, with a minimum of extremes around 1930. Kunkel suggests that as a consequence of this, natural variability cannot be discounted as an important contributor to recent increases in extremes.

Extremes derived from monthly data

Using global monthly gridded datasets for 1900–1988, Dai *et al.* (1997) found, using EOF (empirical orthogonal function[1]) analysis, that ENSO is the single largest cause of extremes in precipitation (as well as the cause of interannual variability), accounting for 15–20% of the global variance of precipitation, while in the ENSO regions themselves (see Figs. 14.6 and 14.7), the contribution is considerably more than 20%. Dai *et al.* also found that severe droughts occurred more frequently in the first half of the twentieth century in central USA and Europe, lessening during the last four to five decades of the century. Many of the severe droughts and floods were also found to be associated with ENSO events. It also emerged from this study that the NAO accounts for about 10% of winter precipitation variance in its region of influence.

Another study of global behaviour using monthly data was based on the GHCN/Hulme (1994) dataset gridded to 2° latitude/longitude resolution by taking

---

[1] EOFs, like principal component analysis (PCA), are statistical methods used to simplify a complex dataset, such as one involving variations in both space and time as in global precipitation, in order to extract a simplified pattern or 'optimal' representation of data with both space and time dependence (Lorenz 1956). It is similar to Fourier analysis, which separates out the individual components of a complex waveform.

a simple average of all stations in each cell (New *et al.* 2001). Only stations that had at least a 25-year run of data were used and this limited the global land areas that could be covered. It was found from this study that the USA, Europe, southern Africa and Australia showed a relatively consistent increase in the frequency of wet spells between the start and end of the twentieth century. Over the Midwestern US, tropical North Africa, the western coast of Australia and the Asian monsoon region there was a reduction in their frequency. Where wet spells had increased in frequency there was an equivalent decrease in the number of dry spells. However, over central South Africa and southeast India, there was an increase in the frequency of both wet and dry spells, while over central Asia there was a tendency for a decrease in the frequency of both.

So is there a clear trend in extremes over a long period? This will not be easy to answer. For this reason, I would particularly like to see a detailed analysis of the long record of precipitation that we have from across the UK going back to 1766 (see Fig. 13.4), to establish the frequency and magnitude of extremes occurring over the whole of this 240-year period. I understand that the data have not yet been compiled in a way that allows extremes back to 1766 to be estimated, although it is planned to do this shortly. This will help to establish whether extreme events are greater now in the UK than two centuries ago or whether extremes have a cyclic nature. This will not tell us anything about global extremes, but it is probably the longest and most extensive record of precipitation available from anywhere on earth and so is quite unique and deserves close scrutiny.

If the foregoing chapters tell us anything certain, however, it is that the present network of raingauges is woefully lacking in precision, time resolution and geographical cover and also that the network is in decline. This prevents us from determining with adequate precision how precipitation is changing. The last chapter looks at how this can be corrected.

References

Allen, R., Lindesay, J. and Parker, D. (1996). *El Niño–Southern Oscillation and Climate Variability.* Melbourne: CSIRO.

Bjerknes, J. (1969). Atmospheric teleconnections from the equatorial Pacific. *Monthly Weather Review*, **97**, 163–172.

Corti, S., Molteni, F. and Palmer, T. N. (1999). Signature of recent climate change in frequencies of natural atmospheric circulation regimes. *Nature*, **398**, 799–802.

Dai, A., Fung, J. and Del Genio, A. (1997). Surface observed global land precipitation variations during 1900–1988. *Journal of Climate*, **11**, 2943–2962.

Groisman, P. Y., Karl, T. R., Easterling, D. R. *et al.* (1999). Changes in the probability of heavy precipitation: important indicators of climate change. *Climate Change*, **42**, 243–283.

Hennessy, K. J., Suppiah, R. and Page, C. M. (1999). Australian rainfall changes, 1910–1995. *Australian Meteorological Magazine*, **48**, 1–13.

Hulme, M. (1994). Validation of large-scale precipitation fields in General Circulation Models. In *Global Precipitation and Climate Change* (ed. M. Desbois and F. Desalmand). Berlin: Springer, pp. 387–406.

Hurrell, J. W. (1995). Decadal trends in the North Atlantic Oscillation: regional temperatures and precipitation. *Science*, **269**, 676–679.

Hurrell, J. W. and van Loon, H. (1997). Decadal variations in climate associated with the North Atlantic Oscillation. *Climate Change*, **36**, 301–326.

Iwashima, T. and Yamamoto, R. (1993). A statistical analysis of extreme events: long term trend of heavy daily precipitation. *Journal of the Meteorological Society of Japan*, **71**, 637–640.

Jones, P. D., Jonsson, T. and Wheeler, D. (1997). Extension to the North Atlantic Oscillation using early instrumental pressure observations from Gibraltar and south-west Iceland. *International Journal of Climatology*, **17**, 1433–1450.

Karl, T. R. and Knight, R. W. (1998). Secular trends of precipitation amount, frequency and intensity in the United States. *Bulletin of the American Meteorological Society*, **79**, 231–241.

Konnen, G. P., Jones, P. D., Kaltofen, M. H. and Allan, R. J. (1998). Pre-1866 extensions of the Southern Oscillation index using early Indonesian and Tahitian meteorological readings. *Journal of Climate*, **11**, 2325–2339.

Kunkel, K. E., Easterling, D. R., Redmond, K. and Hubbard, K. (2003). Temporal variations of extreme precipitation events in the United States: 1895–2000. *Geophysical Research Letters*, **30** (17), 1900.

Lander, M. A. (1994). An exploratory analysis of the relationship between the tropical storm formation in the western Pacific and ENSO. *Monthly Weather Review*, **122**, 636–651.

Lorenz, E. N. (1956). Empirical orthogonal functions and statistical weather prediction. Statistical Forecasting Project, Scientific Report 1. Cambridge, MA: MIT.

New, M., Todd, M., Hulme, M. and Jones, P. (2001). Precipitation measurements and trends in the twentieth century. *International Journal of Climatology*, **21**, 1899–1922.

Osborn, T. J., Hulme, M., Jones, P. D. and Basnett, T. A. (2000). Observed trends in the daily intensity of United Kingdom precipitation. *International Journal of Climatology*, **20**, 347–364.

Philander, S. G. H. (1990). *El Niño, La Niña and the Southern Oscillation*. New York, NY: Academic Press.

Ropelewski, C. F. and Jones, P. D. (1987). An extension of the Tahiti–Darwin Southern Oscillation Index. *Monthly Weather Review*, **115**, 2161–2165.

Serra, Y. L., A'Hearn, P., Freitag, H. P. and McPhaden, J. (2001). ATLAS self siphoning rain gauge error estimates. *Journal of Atmospheric and Oceanic Technology*, **18**, 1989–2002.

Walker, G. T. and Bliss, E. W. (1932). World weather. *Memoirs of the Royal Meteorological Society*, **4**, 53–84.

Weather (1998). El Niño special edition. *Weather*, **53**, 270–336.

Part 5

Future developments

Starting with the deliberations of the Greek philosophers 2500 years ago, but delayed firstly by Roman indifference to the Greek tradition of academic enquiry, followed by the suppression of secular thought by the church through the next 2000 years, then revived by the Enlightenment in Europe and refined by the scientific revolution of the last few hundred years, we have finally arrived at a thorough understanding of the processes leading to precipitation. There are some areas that still need refinement, such as better insight into a number of aspects of lightning, into the growth of snow crystals and the dynamics of some clouds, but otherwise our understanding of the processes involved is reasonably complete.

For the future, new and different tasks have arisen in the wake of these past achievements. Firstly there are still many old raingauge records that have not yet been collected together and assembled into datasets, but there is plenty of vibrant activity in progress compiling, gridding and analysing these data at the world's leading government and academic research centres. The measurement of precipitation, on the other hand, lags far behind.

The first known 'scientific' measurement of rainfall was made by Castelli in 1639 in Italy, followed soon afterwards by Sir Christopher Wren and Robert Hooke, among others, in England. Over the next 450 years improvements were made to raingauges and to how they were exposed, but progress was relatively slow and precipitation measurement has not kept pace with our greatly improved understanding of the processes of precipitation. So much so that even today we are still using many instruments not that dissimilar to those made centuries ago: compare for example Figures 7.1 and 8.1 – the Korean gauge of the fifteenth century and the Mk 2 5-inch from the twentieth. To advance our understanding of precipitation, this imbalance needs correction, for no science progresses without precise and improving measurements, as Aristotle failed to appreciate. Late in the day Leonardo and Bacon came to realise that, as Kepler put it, *to measure is to know*. John Locke also stressed that scientific truth needs evidence from the real world. I am, therefore, devoting the last chapter to exploring how precipitation measurement should be improved so that we keep our feet firmly on the ground.

15

The future of precipitation measurement

Raingauges are the gold standard of precipitation measurement. They are not about to be made redundant by radar, satellites, optical raingauges or disdrometers. They are the key instruments, the principal source of data. However, while not being a substitute for raingauges, satellite passive and active microwave sensors are extremely valuable additional tools, which, when combined with raingauge measurements, offer the most promising way forward. This chapter looks at what needs to be done to get the best out of both.

In considering the future, I will concentrate on the measurements required to obtain accurate global and regional means, trends and variability of precipitation through the twenty-first century along with an estimate of extremes. These measurements are needed not just in order to follow climate change for the knowledge itself, in the spirit of the ancient Greeks, but for the practical tasks of validating climate models, for adjusting satellite and radar algorithms and for calibrating these remote sensor readings. The data are also needed to allow plans to be made to cope with climate change and will be of great practical value, in the spirit of the Roman Empire, to agriculture, water management and flood warning. Trenberth *et al.* (2004) give a good précis of future needs.

Limitations of the existing raingauge network

While there is already a worldwide network of raingauges in operation, on land, it is unsatisfactory, as explained in Chapter 8, because of the wide variety of gauges used and the different methods of exposing them, which introduce a whole train of errors. This old network is mostly operated by the world's National Weather Services (NWSs), overseen by the WMO and established primarily for synoptic use over the last hundred years or so. The sites were selected more through historical accident and practical convenience than on how well they represented an area. The standard of operation of the gauges also varies markedly from one country to the

279

next, and from countries that are engaged in civil war or suffer poverty or disease few data may be forthcoming, thereby generating data-sparse regions. And as the IPCC and the WMO have anxiously pointed out, the number of stations is also in decline. In addition, some data are not shared internationally for political reasons. There can also be decades of delay before data finally filter their way through the bureaucracy of large organisations into the public domain. Add to this the errors of the raingauges themselves, and the end product, the raw data, is inevitably of questionable accuracy and limited geographical extent. Many of the measurements are also at best daily and often only monthly totals, preventing a complete understanding of the processes involved, especially extremes. And this applies just to the land stations; over the 71% of the globe covered by the sea hardly any measurements are made at all, instruments being restricted to just a few islands. In consequence, a new network of raingauges is required, free from these historical limitations, to collect the data required for future climate studies and environmental management.

The old network can continue to operate while the new network is gradually installed, allowing some back-correction of the old data in due course. The new network should not, however, be seen in terms of new raingauges replacing the old ones at existing sites. It would be a new, not a replacement network, offering the chance to start afresh with new instruments at new and better sites under international supervision to worldwide standards.

The new raingauge network

If the new raingauge network is to take full advantage of current knowledge and technology, it is necessary to specify its characteristics carefully and to ensure that the whole system conforms to this. Regional and national differences and foibles, which permeate the current network, must be avoided in future.

The gauges and their exposure

A glance back to Chapter 8 will remind readers that the main raingauge errors are due to their exposure to the wind and to outsplash from shallow gauges, yet these problems are not difficult to overcome. The aerodynamic error is virtually eliminated by placing gauges in pits with their funnels level with the ground (Fig. 8.9) and where this is not possible, wind losses can be reduced, although not eliminated, by using aerodynamically-shaped gauges (Fig 8.10) or wind shields (Fig. 8.6). Outsplash can be cut to negligible proportions by using gauges with sufficiently deep funnels. These simple requirements are easily met, entirely practicable and minimal in cost. Their adoption would greatly improve the quality of the raw data without the need to invoke new, complex or expensive instruments; we have what is needed already to hand.

The most common way to measure the collected rainwater is by tipping bucket, although there are alternatives such as measuring the weight or electrical capacitance of the water as it accumulates in a container. All of these have the potential to be accurate (and inaccurate) but I would opt for 0.1 mm tipping buckets with a funnel area of 1000 cm^2 (to ensure sufficient water for accurate measurement). There is a satisfying and reassuring simplicity, reliability, robustness and cheapness to this method.

The measurements would be logged on-site or telemetered in near-real-time via satellite to a distant base. The time and date of each bucket tip would be recorded to allow precipitation intensity and duration to be calculated, offering an improvement on the daily or monthly totals available from the old network, which masks many important details – daily totals do not tell us anything about rainfall intensity through the day.

Choice of sites

Even when operated in pits, the gauges must still be positioned correctly within the site at which they are operated, being no closer to any object than four times the object's height. And the site itself must also be appropriately chosen to be representative of the surrounding area (see Chapter 8). The number of raingauges I have seen around the world that are badly exposed is considerable. Site conditions would also need to be maintained over decades and checked periodically, the sites and their stability being as important as the instruments.

Metadata

A record of any changes at the sites ('metadata') would need to be kept so that future users of the data could check for sudden, or gradual, changes and thereby attempt to correct for them. Raingauge recalibration and annual photographs of the site would be useful entries in any metadata diary. Such records are lacking at most existing sites, making it difficult for researchers to correct the data. Sites would also need to be chosen that were unlikely to change significantly over a long period.

Geographical cover

The existing land-based raingauge network has grown slowly in an unplanned, ad hoc fashion over the last century, the sites often not being representative of the area. Starting afresh and with hindsight, the establishment of a new network would enable more appropriate sites to be selected, and this would be the first task in setting up a new raingauge network. Key locations would be chosen for a minimal initial network of a few hundred gauges, with additional sites identified for subsequent expansion, up to perhaps two thousand.

While there are a few dozen buoys currently measuring precipitation in the tropical Pacific and Atlantic oceans (see TAO/TRITON in Chapter 14) the use of buoys is expensive and cannot be the general answer in any new network for measuring precipitation over the oceans. Nor are such unstable platforms conducive to accurate precipitation measurement, particularly in the more windswept and rougher seas of the higher latitudes. Ships provide an even less ideal platform because of still more severe exposure problems compared with land sites. Ships also follow a limited set of routes, often modified to avoid bad weather and thereby missing the rain. Their readings also do not refer to one particular location.

While islands have the disadvantage that their presence may modify the local precipitation, gauges installed on them nevertheless give invaluable data. In planning the new network, therefore, full and extensive use should be made of them. An important consideration would be where to place the gauges on the islands to avoid locations where the island influences the rainfall. With modern instruments, however, regular attention is not necessary and the islands could be remote, small and uninhabited. Improved measurements over the oceans would be an important step forward. Covering 71% of the planet's surface, being fairly unfettered by political rankling and lacking the complication of mountains and the tedium of vandalism, oceans offer an ideal location for instruments.

Radar's future is reviewed below, but it is worth jumping forward briefly to speculate upon its use on islands. Placed on high ground, radar would have a view to the horizon over 360 degrees, unobstructed by the uneven topography that hampers its use on land, covering a good-sized area over which the precipitation could be measured. There would be the option to develop software that removed island-induced precipitation. But radar does not give very precise estimates of total rainfall and it would be necessary to have raingauges in the area to help calibrate the radar estimates. Another limitation would be cost, although it might be justifiable at some key locations.

Snowfall measurement

We have to resign ourselves to not being able to measure snowfall very well, at present, it being more difficult to measure than rainfall, mostly because of the much greater effect that wind has on catch. There is no simple solution to this, no equivalent to the pit raingauge. There is the additional problem that dry snow can blow from one place to another after falling and be caught by the gauges.

From Chapter 9 it will be recalled that the available options for measuring snow include exposing the gauge within bushes cut to gauge height or exposing them within a forest clearing. WMO did not select either of these as the 'intercomparison reference' (against which to compare other methods) because they would not be available everywhere and would vary from place to place. Instead WMO opted for the large and expensive Double Fence Intercomparison Reference (DFIR) screen (Fig. 9.2). Tests in Russia had shown that the bush method was best, the DFIR catching 92% of the bush exposure while the smaller screens (Alter etc.) were much less effective.

Although not suitable for use as an international standard, exposure in forest clearings, where available, would be a good option for the new network. Clearings would probably not remain unchanged over a long period but there is no perfect solution to snow measurement.

Where forest exposure is not practicable, the DFIR should be used, despite its cost, at a few important sites. There is scope, however, for further research, and it is possible to visualise a simpler DFIR or simulated 'bushes' as alternatives, although it would be important to avoid repeating the work already carried out very thoroughly over a period of seven years by Huddleston (Chapters 8 and 9) in the 1920s and 1930s in the Lake District. Another line of research would be to evaluate the effectiveness of aerodynamic collectors in catching snow.

Having caught the snow it is then necessary to melt it. Although methods using antifreeze are available, electrical heating is a simpler and cleaner option and also allows tipping buckets to be used to measure the melted snow. To keep raingauge funnels sufficiently warm electrically, however, requires anything from 300 to 1000 watts depending on gauge size, air temperature and windspeed. This will only be possible where mains power is available and so is limited to inhabited regions. Bottled gas is an alternative heat source but needs replenishing and transporting, nor would it be practicable for aerodynamic gauges because the funnel is not surrounded by an outer cylindrical cover.

Since snowfall is only seasonal in many places, gauges would have to be able to measure both snow and rain. Exposure in forest clearings would be effective for both forms of precipitation, because such exposure could be almost as efficient as pit exposures in overcoming the wind problems when measuring rainfall.

It is somewhat more practicable to measure the depth or weight of the snow after it has fallen, and in some circumstances such techniques would be worth including as alternatives to, or in conjunction with, snowfall gauges – depending on the quantity and distribution of snowfall and the depth and permanence of any snowpack. In other situations the measurement of the snow as it melts would offer another possible tool in what is an intractable problem.

Funding and management

Although the WMO is currently committed exclusively to the old instrument network of the NWSs, it might nevertheless still wish to consider supporting or promoting a study to explore the whole matter from the new perspective outlined here. The IPCC should also be concerned and involved, for it is they who have drawn attention to the decline of the existing instrument networks and it is they who urge action over climate change. The World Bank might also wish to be involved.

However funded or organised, any new network would need to be centrally managed so that its performance could be tracked across the globe. There could be no no-go areas or variations in standards from region to region to placate sensibilities, for that would perpetuate the current situation. Management would need to be central and access to all sites by international inspectors would be required. There could be no secrecy about the data, which should be made freely available in one standard format. How this might all be achieved would need discussion, but it should not be beyond the reach of good diplomatic negotiation. If some countries were not able to cooperate then they could be left out of the new network for the moment, but all those who profess to be concerned about the environment can hardly justify the erection of barriers against its measurement. The oceans, as noted above, offer a more benign environment when it comes to human political frailty, and as they cover nearly three-quarters of the globe they are ideal for deploying a new network.

The future of precipitation measurement by radar

The UK is now completely covered by precipitation radar (Fig. 10.1) and in the developed world it is affordable and common. But because of complexity and cost, radar will probably never figure greatly in the future measurement of *global* precipitation. However, in the section above on the oceans, I did suggest that radar might be useful on a selection of islands.

During the coming decades the refinement of algorithms will continue, but there will remain uncertainty over the precision of conversion from reflectivity to precipitation intensity and thence to total amounts; it may not be possible to improve on this beyond a certain limit.

The future of precipitation measurement from satellites

The future of satellites for measuring precipitation lies principally in passive microwave radiometers, supplemented by active precipitation radars, as on the

Tropical Rainfall Measuring Mission (TRMM) satellite. The TRMM satellite will burn up on re-entering the earth's atmosphere around 2010–2012, ending its very extended mission, hopefully to be replaced by the new Global Precipitation Measurement (GPM) mission, with improved sensors, at about the same time. It is also to be hoped that there will be an overlap of the two satellites to allow a comparison of sensor calibrations.

But with the variable nature of precipitation in time and space, combined with the brief snapshot glimpse that satellites give of precipitation just every few hours, as well as their coarse spatial resolution and possible drifts in sensor calibration, satellite measurements are limited and will remain so for some time. It would require the launching of many satellites to give good enough temporal cover. If passive radiometers (with sufficient resolution) could be placed in geostationary orbit, repeat measurements could be made twice an hour.

The place of satellites is, and will probably remain, as an adjunct to a good raingauge network. To operate effectively satellites need good ground truth, and this is a powerful additional argument for the introduction of a new and more accurate raingauge network. Raingauges and satellites together make a powerful combination, and they represent the way forward.

Measuring air temperature

While improving precipitation measurement, we ought also to consider doing the same for air temperature. Global mean temperatures are basic to climate studies, but most temperature data used for this purpose originate from the same NWS sites that measure precipitation, and so share many of their shortcomings. Temperatures measured at these sites come traditionally from Stevenson screens with mercury-in-glass thermometers, and these are prone to many inaccuracies. If and when we embark on the installation of a new raingauge network, it would be invaluable to include new temperature sensors alongside the gauges. To get the accuracy required, aspirated screens would be essential, that is screens that circulate the air actively through the radiation screen over the platinum resistance thermometers by means of a fan, rather than rely on natural ventilation by the wind as Stevenson screens (and most of the new miniature screens) do; poor ventilation leads to quite high errors, especially in bright sunlight. More precise daily means could also be obtained by sampling the temperature every minute instead of the much less accurate daily means obtained by taking the mean of the maximum and minimum thermometer readings, as is typically done at conventional stations.

Concluding thought

We have come to the end of a long journey of enlightenment and discovery covering three thousand years. From time to time, I have had need to mention the delay in progress in understanding the natural world through religious intolerance of secular thought. We think we have now come out of this shadow – it is a long time since Galileo was put under house arrest for life, and Torricelli was silenced – but have we? There is a growing fundamentalism in many religions that is becoming threatening again to science and to free objective thought. Perhaps this is because science comes up with some disturbing and often counter-intuitive discoveries that the majority of people would rather not face up to or perhaps do not understand. But as Richard Dawkins puts it in *A Devil's Chaplain* (2003), 'There is deep refreshment to be had from standing up full-face into the keen wind of understanding.' May we continue to do so.

References

Dawkins, R. (2003). *A Devil's Chaplain*. London: Weidenfeld & Nicolson.
Trenberth, K. E., Moore, B., Karl, R. T. and Nobre, C. (2004). Monitoring and prediction of the earth's climate: a future perspective. Invited paper for the *First International CLIVAR Scientific Conference*, June 2004, Baltimore, MD.

Index